PHYSICS RESEARCH AND TECHNOLOGY

# HELIUM-3

## CRITICAL USES AND SUPPLY CHALLENGES

# PHYSICS RESEARCH AND TECHNOLOGY

Additional books in this series can be found on Nova's website under the Series tab.

Additional E-books in this series can be found on Nova's website under the E-books tab.

PHYSICS RESEARCH AND TECHNOLOGY

# HELIUM-3

## CRITICAL USES AND SUPPLY CHALLENGES

GABRIEL A. ROSS
EDITOR

Nova Science Publishers, Inc.
*New York*

For permission to use material from this book please contact us:
Telephone 631-231-7269; Fax 631-231-8175
Web Site: http://www.novapublishers.com



**Library of Congress Cataloging-in-Publication Data**

Helium-3 : critical uses and supply challenges / editor, Gabriel A. Ross.
p. cm.
Includes index.
ISBN 978-1-61470-823-0 (hardcover)
1. Helium--Isotopes--Industrial applications. I. Ross, Gabriel A.
QD181.H4H44 2011
553'.971--dc23

2011027345

*Published by Nova Science Publishers, Inc. † New York*

# CONTENTS

# PREFACE

The world is experiencing a shortage of helium-3, a rare isotope with applications in homeland security, national security, medicine, industry, and science. For many years the supply of helium-3 from the nuclear weapons program outstripped the demand for helium-3. The demand was small enough that a substantial stockpile of helium-3 accumulated. After the terrorist attacks of September 11, 2001, the federal government began deploying neutron detectors at the U.S. border to help secure the nation against smuggled nuclear and radiological material. The deployment of this equipment created new demand for helium-3. Use of the polarized helium-3 medical imaging technique was increased. As a result, the size of the stockpile shrank. This book discusses the nature of the shortage; federal actions undertaken so far to address it and current and potential sources of helium-3 and options for increasing the supply.

Chapter 1- The world is experiencing a shortage of helium-3, a rare isotope of helium with applications in homeland security, national security, medicine, industry, and science. For many years the supply of helium-3 from the nuclear weapons program outstripped the demand for helium-3. The demand was small enough that a substantial stockpile of helium-3 accumulated. After the terrorist attacks of September 11, 2001, the federal government began deploying neutron detectors at the U.S. border to help secure the nation against smuggled nuclear and radiological material. The deployment of this equipment created new demand for helium-3. Use of the polarized helium-3 medical imaging technique also increased. As a result, the size of the stockpile shrank. After several years of demand exceeding supply, a call for large quantities of helium-3 spurred federal officials to realize that insufficient helium-3 was available to meet the likely future demand.

Chapter 2- Helium-3 gas is a key component of equipment used at ports and border crossings to detect radiation and prevent the smuggling of nuclear material into the United States, among other uses. The National Nuclear Security Administration (NNSA), a separate agency within the Department of Energy (DOE), extracts helium-3 and controls the inventory. Since 2003, NNSA has made helium-3 available for sale to DOE's Isotope Development and Production for Research and Applications Program (Isotope Program). After September 11, 2001, demand increased for radiation detection equipment, and in 2008, the federal government learned that it faced a severe domestic shortage of the gas. GAO was asked to review DOE's management of helium-3 to (1) determine the extent to which the federal government's response to the helium-3 shortage was affected by DOE's management of helium-3; (2) determine the federal government's priorities for allocating the limited supply of helium-3; and (3) describe the steps that the federal government is taking to increase the helium-3 supply and develop alternatives to helium-3. GAO reviewed DOE and NNSA documents and interviewed cognizant agency officials.

Chapter 3- The shortage of He-3 was an inevitable consequence of a declining source from the U.S. nuclear weapons enterprise and a growing demand. However, the crisis and its jarring impacts were avoidable. With foresight on the part of DOE, the kinds of prioritization efforts now happening through the IPC could have started years ago. Research into alternatives to He-3 could have been well along to success, with some areas (such as portal monitor systems) lending themselves to alternatives more readily than others (cryogenics). In short, the stockpile could have been managed in a way that allowed for non-disruptive impacts to industry, researchers and the national security community. Instead, everyone is surprised and scrambling to identify alternatives, suspending their research and their production lines while hoping that a breakthrough in sources of He-3 or alternatives to He-3 happens very, very rapidly. The failure to manage the stockpile with an eye to demand, supply and future needs has had real consequences for many, many fields. Once the shortage became clear to all the key agencies, an interagency process that has laid out a rational guide to allocation and policies has emerged very quickly and appears to be well managed.

Chapter 4- Five months ago, the Subcommittee held a hearing titled: The Science of Security: Lessons Learned in Developing, Testing and Operating Advanced Radiation Monitors. That hearing examined technical problems in the development of the Domestic Nuclear Detection Office's (DNDO's) new generation of radiation portal monitors called Advanced Spectroscopic Portals or ASPs. Among the issues the Subcommittee had expressed an interest in was the impact a reported shortage of Helium-3 was having on the ASP program.

Chapter 5- As requested, my testimony today will address the following points:

- How we became aware of the shortage of He-3;
- How we responded to it;
- What was done at the beginning of the Advanced Spectroscopic Portal (ASP) program to ensure there was an adequate supply of He-3 to meet the program's needs;
- The impact of the shortage on DNDO's radiological and nuclear detection programs; and
- The status of the work we are doing to identify alternative technologies to replace He-3 as a neutron detector.

Chapter 6- Thank you Mr. Chairman, Ranking Member Broun, and Members of the Committee. I appreciate the opportunity to appear before you to provide testimony on the DOE's role and reaction to the national Helium-3 ($^{3}$He) shortage. Both the National Nuclear Security Administration (NNSA), and the DOE Isotope Development and Production for Research and Applications Program (Isotope Program) recently transferred to the Office of Science in the FY 2009 Appropriation, play a role in Helium-3 production and distribution. I have served as the Director of the Office of Science since June 2009, and I am pleased to share with you my perspectives on the role of the DOE Isotope Program in $^{3}$He production and distribution.

Chapter 7- GE Energy's Reuter Stokes has over 50 years of experience supplying radiation detectors. We design and manufacture detectors for Boiling Water Reactors (BWR), neutron scattering instruments, oil and gas exploration, homeland security and nuclear safeguards systems. Our BWR in-core detectors monitor reactor power levels and provide signals to initiate protective actions in the event of an abnormal condition. Our Helium-3 gas-filled neutron detectors are used to accurately account for nuclear materials during handling and processing. Over 35,000 GE Helium-3 detectors are installed in systems deployed around the world today to monitor for the illicit trafficking of smuggled nuclear materials. I look forward to providing you with GE's perspective on the consequences of the Helium-3 supply crisis.

Chapter 8- Chairman Miller, Ranking member Broun, and members of the Committee, my name is Richard Arsenault and I am the Director of Health, Safety, Security and Environment along with being the Corporate Radiation

Safety Officer for ThruBit LLC (ThruBit Logging Solutions) which is a Shell Technology Ventures Fund 1 BV Portfolio company formed in 2005. Today we offer complete logging solutions based on a unique patented "through the bit" deployment technique that provides significant advantages in many applications. We are a small company taking this new technology from proof of concept to commercial introduction with aspirations to grow into a much larger company. I have been involved in the Oil Well Logging industry since 1979 starting out as an Open Hole Wireline Engineer in West Texas and later got involved in the early stages of Logging While Drilling in 1982.

Chapter 9- Mr. Chairman and members of the committee, thank you for the opportunity to testify about the negative impact on scientific research caused by the shortage of helium--three. I am a physics professor at Northwestern University, and I rely heavily on helium--three to carry out scientific research at low temperatures and have been involved in this work since 1970. Low--temperature research is essential for studying properties of materials, such as superconductivity, and magnetism, and for developing various advanced materials. Low--temperature research is also critical to future improvements in metrology and high--speed computation, including quantum information technology. Shortages of helium--three, driven by increased homeland security demands and decreased production capability, are already creating major difficulties in these areas of research.

Chapter 10- If we ask seasoned pulmonologists today how much the practice of pulmonary medicine has changed in the last 25 years, responses will largely be that very little has changed—a few new drugs are available, but there is largely the same technology for measuring lung function and for treatment. $^{3}$He MRI, however, is beginning to emerge as a new "gold standard" and revolutionary biomarker for measuring pulmonary function and structure. Its high signal creates detailed images of lung ventilation and dynamics, and its physical properties allow precise measurement of alveolar size, microstructure, and regional lung function. This makes $^{3}$He MRI particularly sensitive to changes in both global and regional lung function and structure. We are at the cusp of leading pulmonary medicine to a renaissance of new drug development and image-guidance of surgical interventions for various lung diseases, such as asthma, fibrosis, and COPD, which currently affect 11% of the US population. This imaging technology, as I speak, is currently serving as a catalyst for pulmonology to see significant advances in the next 10 years. A lack of supply of $^{3}$He gas will stifle these advances.

In: Helium-3: Critical Uses and Supply Challenges ISBN: 978-1-61470-823-0
Editors: Gabriel A. Ross 

*Chapter 1*

# THE HELIUM-3 SHORTAGE: SUPPLY, DEMAND, AND OPTIONS FOR CONGRESS

***Dana A. Shea and Daniel Morgan***

## SUMMARY

The world is experiencing a shortage of helium-3, a rare isotope of helium with applications in homeland security, national security, medicine, industry, and science. For many years the supply of helium-3 from the nuclear weapons program outstripped the demand for helium-3. The demand was small enough that a substantial stockpile of helium-3 accumulated. After the terrorist attacks of September 11, 2001, the federal government began deploying neutron detectors at the U.S. border to help secure the nation against smuggled nuclear and radiological material. The deployment of this equipment created new demand for helium-3. Use of the polarized helium-3 medical imaging technique also increased. As a result, the size of the stockpile shrank. After several years of demand exceeding supply, a call for large quantities of helium-3 spurred federal officials to realize that insufficient helium-3 was available to meet the likely future demand.

Policymakers now face a number of challenging decisions. In the short term, these decisions are mainly about how to allocate a scarce resource in the face of competing priorities: science versus security, the private sector versus the public sector, and national needs versus international obligations.

Applications with unique needs may pose particular challenges. For example, some types of cryogenic research can only be accomplished using helium-3, whereas in medical imaging and neutron detection, helium 3 has advantages but also alternatives. In the longer term, policymakers also face choices about how or whether to increase helium-3 supply or reduce helium-3 demand and about possible alternative mechanisms for allocating supply. It seems likely that a combination of policy approaches will be necessary.

In addition to the nuclear weapons program, potential sources of helium-3 include tritium produced as a byproduct in commercial heavy-water nuclear reactors; extraction of naturally occurring helium-3 from natural gas or the atmosphere; and production of either tritium or helium-3 using particle accelerators. Until recently, the ready supply of helium-3 from the nuclear weapons program meant that these alternative sources were not considered economic. With the current shortage, this perception may change.

The federal response to the helium-3 shortage began only after the shortage had occurred. Policy was established first by an ad-hoc interagency group formed by the Departments of Energy (DOE), Homeland Security (DHS), and Defense (DOD), and then by an interagency committee established by the National Security Staff. The committee developed a rationing scheme for allocating the available helium-3. Some federal and private-sector users received no allocation or an amount less than they had planned. Several federal agencies are investigating alternative sources of helium-3 and ways to reduce the demand.

Congress is just beginning to grapple with the helium-3 problem. In April 2010, Congress held its first hearing whose main subject was helium-3. So far, congressional attention appears to be focused on oversight of the current situation, how it arose, and the processes currently in place for addressing it. In future hearings and legislation, Congress may address additional issues, such as increasing the helium-3 supply, reducing demand, or changing how supply is allocated.

This report discusses the nature of the shortage; federal actions undertaken so far to address it; current and potential sources of helium-3 and options for increasing the supply; current and projected uses of helium-3 and options for reducing the demand; and options for allocating the supply if it continues to fall short of the demand.

# INTRODUCTION

The world is experiencing a shortage of helium-3, a rare isotope of helium with applications in homeland security, national security, medicine, industry, and science. Federal officials have testified that the shortage is acute and, unless alternatives are found, will affect federal investments in homeland security, scientific research, and other areas. Scientists have expressed concern that the shortage may threaten certain fields of research. This report discusses the nature of the shortage; federal actions undertaken so far to address it; current and potential sources of helium-3 and options for increasing the supply; current and projected uses of helium-3 and options for reducing the demand; and options for allocating the supply if it continues to fall short of demand.

## What Is Helium-3?

Helium-3 is an isotope of helium, an inert, nontoxic, nonradioactive gas. Most helium is helium-4. The natural abundance of helium-3, as a fraction of all helium, is very small: only about 1.37 parts per million.[1] Rather than rely on natural abundance, the federal government manufactures helium-3 through nuclear decay of tritium, a radioactive isotope of hydrogen.

## How Is Helium-3 Made?

By far the most common source of helium-3 in the United States is the U.S. nuclear weapons program, of which it is a byproduct. The federal government produces tritium for use in nuclear warheads. Tritium decays into helium-3.[2] This means that the tritium needs of the nuclear weapons program, not demand for helium-3 itself, determine the amount of helium-3 produced.

## What Is Helium-3 Used for?

Helium-3 has properties that currently make it in high demand. Like all helium, helium-3 is nontoxic. Helium-3 also absorbs neutrons. This property has resulted in its widespread use for neutron detection. Neutron detection is a key component of applications in national and homeland security, industry, and science. For example, the federal government uses radiation portal monitors and other neutron detectors at the U.S. border to prevent smuggling of nuclear and radiological material, and the oil and gas industry uses neutron detectors for well logging.[3]

Another property that has increased demand for helium-3 in recent years is the ability to polarize its nucleus. For example, magnetic resonance imaging (MRI) can take advantage of this property to enable real-time visualization of a patient's lung capacity and capability.[4]

Finally, helium-3 has unique cryogenic properties. Low-temperature physicists use a mixture of helium-3 and helium-4 to achieve temperatures just a few thousandths of a degree above absolute zero (millikelvins). At temperatures below 2.5 millikelvin, helium-3 becomes a superfluid.[5]

## How Do Consumers of Helium-3 Obtain Supplies?

Helium-3 does not trade in the marketplace as many materials do. It is produced as a byproduct of nuclear weapons maintenance and, in the United States, is then accumulated in a stockpile from which supplies are either transferred directly to other agencies or sold publicly at auction. The U.S. producer of helium-3 is the National Nuclear Security Administration of the Department of Energy (DOE). The seller for the public auctions is the DOE Office of Isotope Production and Research.

Although public, the helium-3 auction process has not necessarily been competitive. Most auction sales have been to just two companies: Spectra Gases, Inc., now part of the Linde Group, and GE Reuter Stokes, Inc. Many of the auctions have had just one purchaser. Spectra Gases owns the only facility in the United States that is licensed by the Nuclear Regulatory Commission to remove trace amounts of tritium from raw helium-3, which is necessary for most uses. Despite declining supply and increasing demand, the auction price of helium-3 has been relatively steady, at less than $100 per liter.

Before about 2001, production of helium-3 exceeded consumption. In the past decade consumption has risen rapidly, in part because of the deployment of neutron detectors using helium-3 at the U.S. border to help secure the nation against smuggled nuclear and radiological material. Thus starting in about 2001, and more rapidly since about 2005, the stockpile has been drawn down. By 2009, the government and others recognized that ongoing demand would soon exceed the remaining supply.

In March 2009, the Departments of Energy, Homeland Security, and Defense formed an Integrated Product Team to examine helium-3 supply and demand. The team concluded that further allocations of helium-3 from the federal stockpile would be made only by interagency agreement. In July 2009, the White House National Security Staff established an interagency policy committee (IPC) to investigate helium-3 strategies. In September 2009, the IPC concluded that the auctions should cease, that no further allocations should be made to DHS for radiation portal monitors, and that allocations to agencies and others would henceforth be based on three criteria: whether alternatives to helium-3 exist for the planned application; whether the application increases national or homeland security; and whether the required helium-3 is needed to complete prior investments in infrastructure.

## What Is the Public Policy Problem?

If helium-3 traded in a free marketplace, price could mediate supply and demand, with the available supply of helium-3 being allocated on the basis of the willingness of users to pay the price. Because of the peculiar market arrangement for helium-3, however, supply and demand are effectively disjoined; the price does not reflect the marginal cost of additional supply. In fact, the price of helium-3 appears to be much lower than the likely cost of additional supply, so it is not surprising that demand is exceeding supply. In the face of demand outrunning supply, therefore, the supply of helium-3 is currently being allocated by the interagency policy committee on the basis of criteria derived from an assessment of national needs. As a result, a set of interrelated public policy issues arise: (1) whether it is in the public interest to augment supply, and if so, how best to do so; (2) whether it is in the public interest to curtail demand, and if so, how best to do so; (3) whether the current process for allocating existing supplies of helium-3 is acceptable and in the public interest; and (4) whether, looking to the future, an alternative process for allocating existing helium-3 supplies is warranted—such as a revised set of allocation criteria derived from further assessment of needs and priorities, or a market-based process based on aligning price with marginal cost.

Until 2001, helium-3 production by the nuclear weapons program exceeded the demand, and the program accumulated a stockpile. After the terrorist attacks of September 11, 2001, the federal government began deploying neutron detectors at the U.S. border to help secure the nation against smuggled nuclear and radiological material. The deployment of this equipment created new federal demand for helium-3. Use of the polarized helium-3 medical imaging technique also increased. Annual demand for helium-3 quickly exceeded the annual supply. As a result, the stockpile shrank. After several years of demand exceeding supply, federal officials realized that insufficient helium-3 was available to meet the likely future demand. The federal government and others have projected that future demand for helium-3 will significantly increase in future years, causing the disparity between supply and demand to grow.

Other sources of helium-3 exist, both domestically and internationally. Drawing on these sources might require the development of new technologies and approaches, as well as addressing potential international export control concerns. The magnitude of the alternative sources and the ease with which helium-3 could be extracted is unclear. Similarly, demand for helium-3 might be reduced through rationing or the development of alternative technologies, but these alternatives are not yet readily available.

No single policy approach will likely yield sufficient additional supply, reduce predicted demand, or successfully allocate helium-3 to all who wish to have it. Policymakers may have to use a combination of supply- and demand-based approaches to address the helium-3 shortage.

Policymakers now face a number of challenging decisions. Currently, these decisions are mainly about how to allocate a scarce resource in the face of competing priorities: science versus security, the private sector versus the public sector, and national needs versus international obligations. Applications with unique needs may pose particular challenges. For example, some types of cryogenic research can only be accomplished using helium-3, whereas in medical imaging and neutron detection, helium-3 has advantages but also alternatives. To address future helium-3 concerns, policymakers also face choices about how or whether to increase helium-3 supply or reduce helium-3 demand and about possible alternative processes for allocating the supply.

## FEDERAL RESPONSE

While the events of 2001 increased the profile and use of helium-3, the federal government did not foresee the impending shortage. Thus, the executive branch response to the helium-3 shortage began after the shortage had occurred. First an ad-hoc interagency group formed by the Departments of Energy (DOE), Homeland Security (DHS), and Defense (DOD) set policy, then that role passed to an interagency policy committee established by the

National Security Staff. The interagency policy committee developed a rationing scheme for allocating the available helium-3. As a result, some federal and private-sector users received no allocation or an amount less than they had previously planned. Several federal agencies are investigating alternative sources of helium-3 and ways to reduce the demand.

## Identification of the Problem

The DOE, the agency responsible for manufacture and disbursement of helium-3, did not predict the imminent shortage. Neither did any other federal agency that used helium-3. Until 2008, DOE released helium-3 from the stockpile through public auction and interagency transfers, apparently without anticipation of likely future demand. Even though the outflow of helium-3 from the stockpile had become notably greater than the inflow, DOE disbursed increasing amounts each year from 2002 through 2008. In 2008, the call for a large disbursement of helium-3 to the Spallation Neutron Source (a DOE research facility) apparently coincided with a recognition that the helium-3 stockpile could be exhausted and that the available supply might not meet future demand. At a congressional hearing in November 2009, officials of the Domestic Nuclear Detection Office (DNDO, an agency in DHS) testified about the shortage of helium-3 and described the efforts of the federal government to address the shortfall.[6] Before that date, budget justification documents, published research and development strategies, and other agency planning documents did not identify an impending helium-3 shortage, even though they discussed the need for continued investment in neutron detection technology to improve sensitivity and other performance characteristics.

## Executive Branch Actions

In March 2009, DOE, DHS, and DOD formed an interagency Integrated Product Team (IPT) to examine helium-3 supply and demand and related technology R&D. The IPT determined that further allocations of helium-3 would be made only by interagency agreement.[7]

In July 2009, the White House National Security Staff established an interagency policy committee (IPC), with broad federal representation, to investigate strategies for decreasing helium-3 demand and increasing helium-3 supply and to make recommendations about allocation of the existing supply.[8]

In September 2009, the IPC determined that DHS would receive no additional allocation of helium-3 for use in radiation portal monitors.[9] The IPC also determined that auctions of helium-3 held in the stockpile would cease. Instead of auctions, the IPC would allocate the available helium-3 to agencies and others based on three criteria: whether alternatives to helium-3 exist for the planned application; whether the application increases national or homeland security; and whether the required helium-3 is needed to complete prior investments in infrastructure.

The consequence of such a prioritization process is necessarily that lower-priority activities will not have their helium-3 needs met. For example, DHS has had to delay the deployment schedule for the Radiation Portal Monitor program. Some small users, such as academic researchers, have seen the price of helium-3 increase dramatically because of its scarcity.[10] Some international obligations and agreements have not been fulfilled due to these allocation decisions.[11]

Federal agencies have ttempted to locate and recover any unused or surplus helium-3. Some have issued requests for information and proposals regarding the recovery of helium-3 from defunct test equipment.[12] Others have issued similar notices regarding storage and maintenance of agency-based helium-3 supplies.[13]

In addition, federal agencies have accelerated the development and testing of technological alternatives to helium-3 for homeland security purposes. Much of this activity has taken place through the DOE national laboratories. Successful development and implementation of alternative technologies may lead to reduced helium-3 demand.

## Congressional Actions

Congress is just beginning to grapple with the helium-3 problem. In April 2010, Congress held its first hearing focused mainly on helium-3.[14] So far, congressional attention has been focused on oversight of the current situation, how it arose, and the processes currently in place for addressing it. In future hearings and legislation, Congress may address additional issues, such as increasing the helium-3 supply, reducing demand, or modifying the allocation of the limited supply.

# HELIUM-3 SUPPLY

At present, helium-3 is only produced as a byproduct of the manufacture and purification of tritium for use in nuclear weapons. The supply of helium-3 therefore derives mostly, perhaps entirely, from two sources: the U.S. and Russian governments.[15] Other potential sources of helium-3 do exist, but using these sources would present varying degrees of technical and policy challenges. Congress has several options for increasing the supply of helium-3, either from conventional sources or by encouraging the development of new sources. Among the important characteristics of all these potential sources are their likely cost and the amount of helium-3 they could potentially supply. The potential annual production of helium-3 from alternative sources is uncertain. This uncertainty results from incomplete characterization of the sources, variability in helium-3 content, and other factors, such as the willingness of public or private entities to invest in infrastructure to enable production at a particular scale. Even for potentially large sources, producing helium-3 from these sources may be impractical on cost grounds.

## Current Supply

The main source of helium-3 in the United States is the federal government's nuclear weapons program. For many years, the National Nuclear Security Administration (NNSA) and its predecessor agencies have produced tritium for use in nuclear warheads.[16] Over time, tritium decays into helium-3 and must be replaced to maintain warhead effectiveness. The NNSA recycles the mixture of tritium and helium-3 that results from this decay process and reuses the resulting pure tritium. From the perspective of the weapons program, the extracted helium-3 is a byproduct of maintaining the purity of the tritium supply. This means that the tritium needs of the weapons program, not the demand for helium-3 itself, determine the amount of helium-3 produced.

Until 2001, helium-3 production by the weapons program exceeded demand, and the program accumulated a stockpile. To recoup some of the cost of purifying recycled tritium, the program transferred helium-3 from the stockpile to the DOE Office of Isotope Production and Research for sale at auction. Despite these sales, the helium-3 stockpile grew from roughly 140,000 liters in 1990 to roughly 235,000 liters in 2001.[17] Since 2001, however, helium-3 demand has exceeded production. By 2010, the increased demand had reduced the stockpile to roughly 50,000 liters. See Figure 1. Note that these amounts do not account for helium-3 imports and exports, or helium-3 supplies held by other agencies or the private sector.

The U.S. weapons program currently produces tritium by irradiating lithium in a light-water nuclear reactor. Before 1988, the program used heavy-water reactors at the DOE Savannah River Site in South Carolina. In 1988, the last operating Savannah River Site reactor, the K reactor, was shut down for safety reasons. For the next several years, reductions in the nuclear weapons stockpile meant that tritium recycling met the weapons program's needs without additional tritium production.[18] Over time, as the tritium produced before 1988 decayed into helium-3, the total amount of remaining tritium decreased. The annual rate of helium-3 production from the remaining tritium declined commensurately.

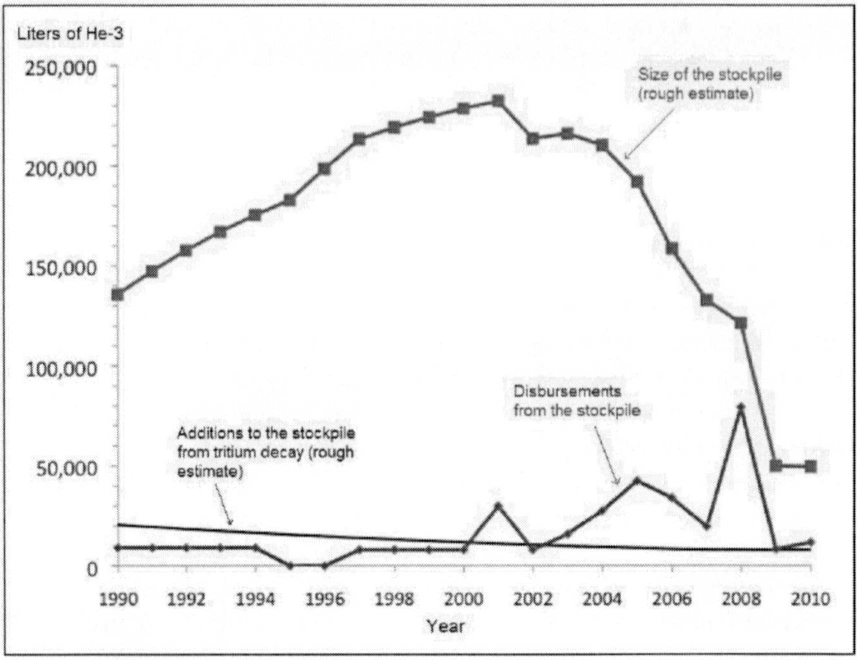


Source: Adapted from Steve Fetter, Office of Science and Technology Policy, "Overview of Helium-3 Supply and Demand," presentation at the American Association for the Advancement of Science Workshop on Helium-3, April 6, 2010.

Figure 1. Size of the Helium-3 Stockpile, 1990-2010.

The DOE restarted tritium production for the weapons program in 2003 using the commercial Watts Bar reactor in Tennessee, operated by the Tennessee Valley Authority (TVA). The tritium production process used there is new and involves irradiation of lithium-containing Tritium-Producing Burnable Absorber Rods (TPBARs). The DOE also built the Tritium Extraction Facility to extract tritium from irradiated TPBARs. This facility, located at the Savannah River Site, became operational in 2007.[19] The NNSA plans to begin irradiating TPBARs at the TVA's two Sequoyah reactors in FY2012.[20] Currently, TVA irradiates only a small number of TPBARs, and tritium production is limited.

As shown in **Figure 1**, the decay of tritium held by the U.S. nuclear weapons program currently generates approximately 8,000 liters of new helium-3 per year. Historically, the price of helium-3 has been $100 to $200 per liter, with commercial prices rising to $2,000 per liter or more after the discovery of the shortage.[21]

## Potential Additional Sources

Potential additional sources of helium-3 include increased production of tritium in light-water nuclear reactors (beyond the amount already produced for the weapons program); extraction of tritium produced as a byproduct in commercial heavy-water nuclear reactors; production of either tritium or helium-3 using particle accelerators; and extraction of naturally occurring helium-3 from natural gas or the atmosphere. Until recently, the ready supply of helium-3 from the nuclear weapons program meant that these alternative sources were not considered economically attractive. With the current shortage, this consideration may change.

### *Tritium from Light-Water Nuclear Reactors*

The production of additional tritium at the TVA reactors (or by similar means at other reactors) would increase helium-3 production correspondingly. The Watts Bar reactor currently irradiates 240 TPBARs per refueling cycle. The Nuclear Regulatory Commission has generically approved the use of approximately 2,300 TPBARs per reactor per cycle.[22] If this capacity were fully utilized at Watts Bar and Sequoyah, annual helium-3 production from the resulting additional tritium could eventually increase by approximately 25,000 liters.[23] At present, however, the site-specific licenses for Watts Bar and Sequoyah permit irradiation of only a much smaller number of TPBARs, and the TVA has limited that number still further because of an issue with tritium permeation into the reactor cooling water.[24] If the current rate of 240 TPBARs per cycle at Watts Bar were successfully extended to both Sequoyah reactors, the result might be an eventual increase in annual helium-3 production of nearly 3,000 liters.[25]

To realize these annual production rates, however, the increased tritium production would need to continue for many years. Because of the slow decay of tritium to helium-3, the production of helium-3 would increase gradually over several decades. Only after more than a decade of increased tritium production would the rate of additional helium-3 production approach even 50% of the projected long-term level.[26]

At present, the nuclear weapons program, in effect, subsidizes the cost of production for helium-3, because the program must manufacture and purify tritium for the ongoing maintenance of its nuclear weapons. The same would presumably be true for additional helium-3 that resulted from increased tritium production to meet the weapons program's needs. If, however, the weapons program increased tritium production to more than the amount it needed, solely in order to generate additional helium-3, it might pass the additional infrastructure and processing costs on to the users of helium-3. According to one estimate, the unsubsidized cost of manufacturing tritium for the nuclear weapons program is between $84,000 and $130,000 per gram.[27] This corresponds to between $11,000 and $18,000 per liter of eventual helium-3.

### *Tritium from Heavy-Water Nuclear Reactors*

Some nuclear reactors, primarily foreign reactors built on the Canada Deuterium Uranium (CANDU) design, use heavy water as a moderating material. Over time, neutrons produced by these reactors convert some of the deuterium in the heavy water into tritium. Some reactor operators, concerned about radiation exposure, remove the tritium from the heavy water when it exceeds a certain concentration. There are currently no commercial heavy-water reactors in the United States, but some foreign facilities have tritium stockpiles that reflect multiple years of operation and separation. Because tritium decays into helium-3, these foreign tritium stockpiles are a potential source of helium-3. Separation of helium-3 from this tritium, while a known industrial procedure, would require specialized facilities and equipment, permits from appropriate government authorities, and perhaps operational changes.

Most of the world's CANDU reactors are located in Canada. Other countries with one or more such reactors include Argentina, China, India,

Pakistan, Romania, and South Korea.[28] The international nature of the CANDU reactor operators may provide additional challenges to using them as a helium-3 source. Each government would likely need to approve the transfer of tritium, a dual-use material, to the United States.[29]

Some of the tritium produced in past operation of heavy-water reactors has already decayed into helium-3 according to its 12.3-year half-life. That helium-3 has probably not yet been separated from the stored tritium, and it seems likely that some or all of it remains available. If so, that helium-3 could provide a larger supply in the near term than the annual production available in subsequent years from continuing decay of the stored tritium. According to one estimate, the tritium stored at Ontario Power Generation in Canada could produce 130,000 liters of helium-3 over ten years: 20,000 liters per year for the first three years and 10,000 liters per year for the next seven.[30] The amount potentially available from other heavy-water reactors is unknown.

The cost of helium-3 from this source is unknown, because it would depend on the price charged for tritium by the foreign supplier. There would also be some costs for separating the helium-3 from the tritium and for storing the remaining tritium as it decayed.

### *Tritium or Helium-3 from Particle Accelerators*

During the 1990s, the U.S. nuclear weapons program considered alternative approaches to producing tritium, especially irradiating lithium with a linear accelerator rather than in a nuclear reactor. In 1998, DOE decided not to proceed with accelerator production of tritium, except as a possible backup option, after concluding that it would be more costly and less flexible than producing tritium in a light-water reactor. Given that decision and the government's subsequent investments in tritium production infrastructure, it seems unlikely that accelerator production of tritium could be an advantageous source of helium-3.

In principle, helium-3 could also be produced directly in an accelerator process, without intermediate production of tritium. Preliminary estimates suggest, however, that this process is impractical on cost grounds.[31]

### *Helium-3 from Natural Gas*

Natural gas reservoirs typically contain impurities as well as the primary component of natural gas, methane. In some cases, these impurities include significant amounts of helium (up to several percent). Suppliers of natural gas often extract this helium in order to increase the energy content of their natural gas and improve its combustion. When a reservoir is relatively helium-rich, it can be economic to purify the extracted helium and sell it as a commodity. In fact, natural gas is the primary commercial source of helium. Domestic natural gas producers extract approximately 80 billion liters of helium each year.[32]

Since 1960, the federal government has maintained a stockpile of raw (unpurified) helium at a facility near Amarillo, Texas.[33] The original purpose of the stockpile was to ensure the availability of helium for national security uses. In the Helium Privatization Act of 1996 (P.L. 104-273), Congress mandated the sale of all but a small portion of the stockpile by 2015. At the end of FY2009, however, more than 500 billion liters of helium remained in the stockpile.[34]

Helium extracted from natural gas, including helium stored in the national helium stockpile, consists mostly of helium-4 but also includes a small proportion of helium-3. The natural gas industry has not historically separated the helium-3 from the helium-4 because, until recently, the federal supply of helium-3 was perceived to be already greater than the likely demand.

An important cost consideration is that some of the processes required to extract helium-3 from natural gas are already undertaken in the production of natural gas and commodity helium. Helium-containing natural gas is purified by liquefaction—cooling it to a temperature at which the natural gas becomes liquid but the helium remains a gas. The helium is separated and then purified by further liquefaction—cooling to a still lower temperature at which the impurities become liquid. The most likely processes for separating helium-3 from helium-4 take place at even lower temperatures, so the fact that helium produced from natural gas is already very cold becomes an important cost advantage. If separation of helium-3 from natural gas took place in conjunction with other natural gas processing, much of the energy required for cooling, and much of the cost of infrastructure and equipment for liquefaction and separation, would already be built into the cost of processing the natural gas.

Separation of helium-3 from helium-4 has been demonstrated on a laboratory scale.[35] Public or private investment in process engineering and development would likely be needed before moving to full-scale production. The amount of helium-3 that could be extracted on a large scale would depend on several factors: access to helium supplies, the proportion of helium-3 in the source helium, the capacity of the processing equipment, and the efficiency of the extraction process.

The U.S. Geological Survey estimates total U.S. helium reserves and resources to be 20.6 trillion liters.[36] Natural gas reservoirs vary in the proportion of helium-3 they contain. A study conducted in 1990 by the

Department of the Interior found ratios of helium-3 to helium-4 that ranged from 70 to 242 parts per billion.[37] These figures imply U.S. helium-3 reserves and resources of between 1 and 5 million liters.[38]

Because of the factors discussed above, any cost estimate for helium-3 extraction from natural gas is inexact. According to one estimate, the energy cost, not including the cost of infrastructure and equipment, might be about $12,000 per liter.[39] Most of this cost, however, would be to separate the commodity helium from the natural gas. Starting with cooled commodity helium, in conjunction with regular natural gas processing, the incremental energy cost of separating out the helium-3 might be $300 per liter.[40] Another source estimates $34 per liter, again excluding the cost of infrastructure and equipment.[41] The difference between these two estimates appears to be their different assumptions about heat exchange efficiency, an issue whose resolution may require development of a prototype processing system. Over and above these energy costs, the cost of a helum-3 extraction plant is estimated to be tens of millions of dollars.[42]

Although U.S. helium-3 reserves and resources are large, the rate at which refiners already extract commodity helium from natural gas limits the amount of helium-3 that could be available per year at the lower cost range ($34 or $300 per liter plus infrastructure and equipment). According to one expert, separating helium-3 from all domestically produced helium would make available about 26,000 liters of helium-3 per year.[43] Producing more than this amount would draw on natural gas that would not otherwise be processed to extract commodity helium, and as a result, the higher cost estimate ($12,000 per liter plus infrastructure and equipment) would apply.

### *Helium-3 from the Atmosphere*

Like natural gas, the atmosphere contains a small quantity of helium, a fraction of which is helium-3. Liquefaction and other processes similar to those used to extract helium from natural gas could similarly be employed to extract helium from air. The processes for separating helium-3 from the resulting mixed helium would also be much the same.

The atmosphere contains approximately 280 billion liters of helium-3. While this makes air by far the largest potential source of helium-3, extracting helium-3 from air is very unlikely to be cost-competitive with extracting helium-3 from natural gas. The concentration of helium in the atmosphere is only about 5 parts per million.[44] As a result, the energy cost of producing helium from air would be about 1,000 times more than that of producing helium from natural gas.[45] The fraction of helium-3 in atmospheric helium is 5 to 20 times more than the fraction of helium-3 in helium from natural gas, but this is not enough to offset the much larger volume of air that would need to be processed to extract a given quantity of helium. Moreover, the cost of producing helium-3 from air would have to include the full energy cost of liquefying the air, whereas the energy cost of liquefying natural gas is already borne by natural gas refiners as part of their regular production process.

## Options to Increase Supply

The helium-3 shortage is a disequilibrium between supply and demand. One way for policymakers to address the shortage, therefore, would be to increase the production of helium-3. The preceding section addressed several possible sources that could be tapped. Increasing the supply could potentially be accomplished in several ways: by increasing U.S. tritium production from current or new sources, importing tritium, importing more helium-3, increasing helium-3 recycling, or producing helium-3 from alternative sources. Congress has a variety of options that might result in these outcomes.

### *Increase Domestic Tritium Production*

Congress could direct NNSA to increase its tritium production beyond the amount needed for nuclear weapons purposes to the amount needed to meet helium-3 demand. Alternatively, Congress could direct DOE or another agency to use a similar technique at other commercial reactors, or it could encourage the private sector to do so on its own by offering subsidies, tax incentives, changes in licensing requirements, purchase guarantees for the resulting tritium production, or other measures. Finally, Congress could decide to reexamine DOE's decision in the 1990s to reject accelerator production of tritium, such as by funding additional research and development or new studies that specifically address the context of the helium-3 shortage. An increased supply of tritium from any of these options would result, in time, in an increased supply of helium-3. Such an approach has several challenges, including timeliness, capacity, cost, and nonproliferation concerns.

#### Timeliness

Even if the federal government increased domestic tritium production immediately, the helium-3 supply would not increase immediately for two reasons. First, the TPBARs used for tritium manufacture are irradiated for a full reactor refueling cycle, which is typically about 18 months.[46] More importantly, once additional tritium is produced, only about 5.5% of it decays into helium-3 each year because of its 12.3-year half-life. These two factors combine to significantly impede domestic production of tritium as a short-term solution. Years of additional tritium production would need to occur before it would yield a sizeable annual supply of additional helium-3.

### Production and Extraction Capacity

Current circumstances would make it difficult to increase domestic tritium production drastically. The capacities and licenses of the Watts Bar and Sequoyah reactors limit their capacity to manufacture additional tritium. With the proper licensing and other arrangements, other commercial light-water reactors could employ the same technique. Incentives would likely be necessary, however, because of the operational changes, such as increased uranium enrichment in the reactor fuel, that are needed to accommodate TPBARs. In 1997, when DOE requested proposals for the tritium production contract, the TVA was the only operator to submit a responsive bid. Even if these challenges could be overcome, NNSA has not yet been able to overcome the technical problems that currently constrain tritium production at Watts Bar and Sequoyah, and according to GAO, it may not be able to produce even the amount of tritium needed for the nuclear weapons program.[47]

The capacity of the tritium extraction facility could also be a limiting factor for increasing tritium production in commercial reactors. If necessary, the capacity of the existing extraction facility could be expanded, or an additional facility could be built. The cost of the existing tritium extraction facility was $506 million.[48]

### Cost

The cost of tritium produced for the weapons program is subsidized by that program. As noted above, the cost of tritium in excess of the weapons program's needs might not be similarly subsidized, in which case it would likely be much higher than the historical helium-3 price.

### Nonproliferation Concerns

Because tritium is a component of nuclear weapons, it is possible that increasing tritium production would raise nuclear nonproliferation concerns. Existing arms control treaties limit the number of U.S. nuclear warheads, not the tritium that is used in them. Nevertheless, some countries or activists might perceive an increased tritium production capacity as enabling or increasing the likelihood of future increases in nuclear warhead capacity. Using civilian facilities to produce tritium might be seen by these observers as a precedent for blurring the separation between civilian and military nuclear facilities; such separation is often an element of U.S. nonproliferation policy in other countries. If U.S. commercial power plants involved in tritium production became perceived as part of the weapons program, some foreign suppliers might be less willing to provide them with components and equipment.

### *Import Tritium*

Another approach to increasing the tritium supply, and thus the future supply of helium-3, would be for the U.S. government to purchase tritium from international sources. Such an approach has limitations, in general because tritium is export-controlled and more specifically because some nations have policies that prohibit the export of materials used in nuclear weapons.[49]

It might be possible to assure supplier countries of non-weapons use by sequestering the imported tritium and using that supply exclusively for production of helium-3. Extracting and purifying helium-3 from this sequestered tritium supply, however, might require separate apparatus and infrastructure. If so, developing and maintaining that duplicate capacity could have substantial costs and would likely increase the price of the resulting helium-3.

One possible source of imported tritium is foreign heavy-water reactors, such as the CANDU reactors used in Canada, India, and elsewhere. International export control agreements would regulate any shipment of tritium from those countries to the United States. Diplomatic efforts or new international agreements might be needed to facilitate changes in these restrictions. Another possible source is Russia, which produces tritium for use in its own weapons program. Russia, however, has a history of exporting helium-3 directly, which may make Russian tritium exports unlikely.

Timeliness would also likely be an issue. Like domestically produced tritium, imported tritium would only decay gradually into helium-3. A one-time supply of imported tritium would take decades to convert into helium-3. Production of helium-3 from a regular annual supply of imported tritium would take decades to reach its long-term potential level.

### *Increase Domestic Recycling of Helium-3*

Because helium-3 is stable and chemically inert, it can be reused indefinitely. Losses through leakage into the atmosphere are essentially unrecoverable, but they are likely to be small in most applications, especially considering the currently high cost of replacing lost helium-3. In some applications, such as medical imaging, recovered helium-3 may need to be purified before reuse. Some opportunities remain, however. For example, Savannah River National Laboratory is developing a process to extract helium-3 from tritium equipment that would otherwise have been discarded; this process may ultimately provide as much as 10,000 liters of additional helium-3.[50] To increase the domestic supply, Congress could mandate or facilitate recycling of existing helium-3. The need for congressional action in this area is uncertain, however, because of the incentive to reuse and recycle that the high market price already provides.

### *Import Helium-3*

Because other countries have supplies of tritium, they either have or could have supplies of helium-3 arising from tritium decay. Between 2004 and 2008, for example, the United States imported approximately 25,000 liters of Russian helium-3 per year.[51] Russia has suspended exports of helium-3, but it might be possible to increase imports from elsewhere. Because helium-3 can be converted back into tritium, it too is generally export-controlled. Some of the export-control challenges discussed above for importing tritium therefore also apply to importing helium-3. However, because imported helium-3 would not need additional separation or purification, a requirement to sequester imported helium-3 would be less burdensome.

Considering the currently high market price of helium-3, it appears likely that any country with the infrastructure needed to extract helium-3 from tritium would already be doing so. The absence of such an increased international helium-3 supply suggests that most countries have little or no helium-3 separation capacity. One way to facilitate increased helium-3 imports would therefore be to help develop helium-3 separation infrastructure in countries that have heavy-water reactors or other tritium sources. Such assistance could take many forms, from providing financial incentives or technical help to establishing a joint venture or providing equipment directly.

If the United States and another country decided to establish an international partnership to produce helium-3, determining an equitable financing mechanism might be challenging. The market price of helium-3 is currently much higher than it has been historically. If demand continues to outstrip supply, developing additional sources may continue to provide a valuable return. On the other hand, if demand falls, perhaps because the high price drives users to identify alternatives, then the return on investments in helium-3 production infrastructure may be less attractive. This uncertainty could be particularly problematic for potential international partners if the United States simultaneously takes policy actions specifically aimed at reducing demand.

### *Produce Helium-3 from Alternative Sources*

Congress could take a variety of actions to encourage the development of alternative helium-3 sources, such as extraction from natural gas and the other techniques described previously in this report. Options for policymakers include funding research and development on new production technologies, funding pilot projects to demonstrate practicality, providing tax incentives to producers, and guaranteeing a minimum future price or a minimum level of future government purchases. Because these alternative sources are less technologically mature than the existing infrastructure, the available options for encouraging their development may be promising only in the long term.

## HELIUM-3 DEMAND

The demand for helium-3 has increased dramatically since 2001. Prior to 2001, the demand was approximately 8,000 liters per year, which was less than the new supply from tritium decay. After 2001, the demand increased, reaching approximately 80,000 liters in 2008. Projections show demand continuing at above the available new supply for at least the next several years. See Figure 2. These projections contain many variables and therefore considerable uncertainty. Some estimates project much higher non-governmental demand, perhaps more than 100,000 liters in FY2011 and FY2012.[52] Some estimates appear to measure helium-3 quantities at nonstandard pressures. Because liters are a volume measure, and all gases change volume depending on their pressure, inconsistency in measurement has the potential to create confusion when amounts projected by different analysts are added. Perhaps most important, given such a large mismatch between supply and demand, users are likely to seek out alternative technologies, reschedule planned projects, and make other changes that reduce demand below what it would be in the absence of a shortage. It is unclear whether the available estimates reflect (or indeed, could reflect) these likely changes. Similarly, it is unclear whether federal agencies and the private sector can reduce demand sufficiently to match the current helium-3 supply and still meet priorities for security, science, and other applications.

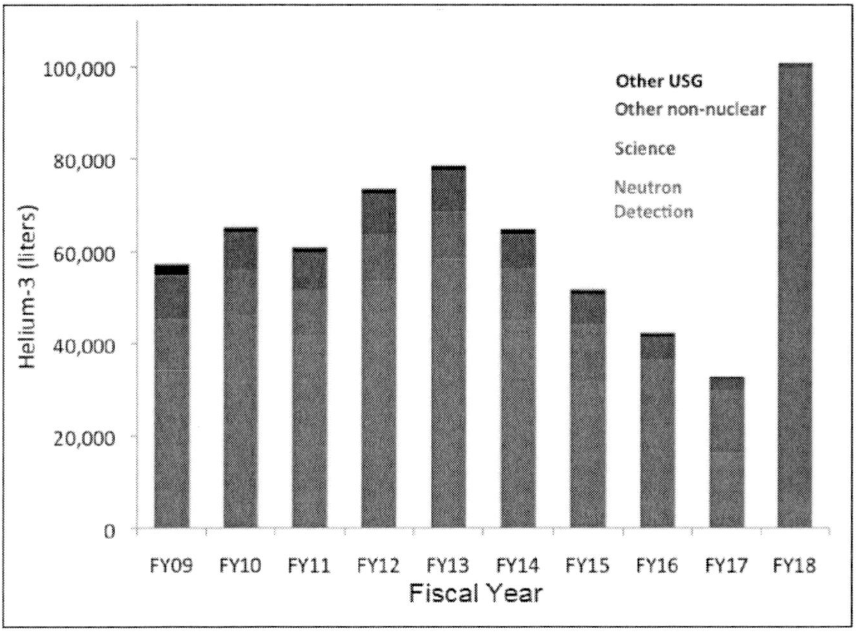


Source: Adapted from Steve Fetter, Office of Science and Technology Policy, "Overview of Helium-3 Supply and Demand," presentation at the American Association for the Advancement of Science Workshop on Helium-3, April 6, 2010.

Note: USG = U.S. Government. Science includes both research activities and large laboratory facilities. Another allotment to the Spallation Neutron Source underlies the increase in scientific demand in FY2018.

Figure 2. Projected Helium-3 Demand, FY2009-FY2018.

## Current Uses of Helium-3

Neutron detection applications in national and homeland security are the largest users of helium-3, but scientific research, medicine, and industry are also significant.

### *National and Homeland Security*

The demand for helium-3 for national and homeland security purposes falls into two main categories: the detection of smuggled radiological and special nuclear material and the monitoring of known special nuclear material to ensure its security.[53]

The Department of Defense, Department of State, NNSA, and DHS all have deployed radiation detection equipment to detect smuggled radiological and nuclear material.[54] Through programs such as Cooperative Threat Reduction, the Second Line of Defense, and the Radiation Portal Monitor program, these agencies have deployed thousands of radiation portal monitors both domestically and overseas. Each portal uses approximately 50 liters of helium-3 as the basis for its neutron detection capability. Some of the programs have been in place since before 2001. Others, such as those operated through DHS, were established later. The broad expansion of these deployments has provided the greatest demand for helium-3 and been the largest drain on the helium-3 stockpile.

The Department of Defense and NNSA also use helium-3 in neutron detectors to ensure that stores of special nuclear material are fully accounted for. Accurate neutron counting over long time periods is one way to monitor the continued presence of materials such as plutonium. In addition, the United States contributes helium-3 to meet the nuclear security and monitoring needs of the International Atomic Energy Agency (IAEA).

Department of Defense guidance and navigation systems for munitions, missiles, aircraft, and surface vehicles include ring laser gyroscopes that use helium-3. Testing and qualification are under way on an alternative gas for this purpose.[55]

### *Science*

Scientific uses of helium-3 are diverse, ranging from neutron detection to cryogenics, laser physics, and research on the properties of helium-3 itself. Large-scale government research facilities, such as the DOE Spallation Neutron Source at Oak Ridge National Laboratory in Tennessee, may use tens of thousands of liters. Numerous smaller, laboratory-scale users are in academia and elsewhere. Although individual university researchers use smaller quantities, their ability to pay the currently high price for market-rate helium-3 may be limited. Although they are part of the private sector, their research is often funded by federal agencies. The United States also participates in international science projects that require helium-3, such as ITER (originally called the International Thermonuclear Experimental Reactor), currently under construction in France.

The United States, as well as engaging directly in scientific activities that require helium-3, has historically been a major source of helium-3 for the international scientific community. Some foreign scientists may depend on U.S.-supplied helium-3 for their research. Although some of these foreign scientists may work independently, many of them are likely colleagues and collaborators of U.S. scientists.

### *Medicine*

The development of a polarized medical imaging technique contributed to an expansion of demand for helium-3 in the late 1990s. This technique depends on patients inhaling polarized gas so that imaging of the gas using MRI provides a visualization of lung function.

### *Industry*

In industrial applications, a neutron-emitting material is coupled with a helium-3 neutron detector to make density measurements based on the number of neutrons reflected back to the detector. This technique is used for oil and gas well logging[56] and to determine the density of road construction.

## Technological Alternatives to Reduce Demand

One way to address the helium-3 shortage would be to reduce demand by moving helium-3 users to alternative technologies. For the largest application, neutron detection, the DOE national laboratories and others have examined numerous alternatives. As discussed below, some technologies appear promising, though implementation would likely present technical challenges. For other applications, alternative technologies may not currently exist.

### *Alternative Technologies for Neutron Detection*

Because of its detection performance, nontoxicity, and ease of use, helium-3 has become the material of choice for neutron detection. Nevertheless, other materials also have a long history of use. With the current shortage of helium-3, researchers are reexamining past alternatives and investigating new ones. Existing alternative neutron detection technologies have significant drawbacks relative to helium-3, such as toxicity or reduced sensitivity. A drop-in replacement technology does not currently exist.[57] The alternatives with most short-term promise as helium-3 replacements are boron trifluoride, boron-lined tubes, lithium-loaded glass fibers, and scintillator-coated plastic fibers. A new scintillating crystal composed of cesium-lithium-yttrium-chloride (CLYC) also appears promising. Other materials, less suitable in the short term, show promise for the long term. Before the helium-3 shortage became apparent, most neutron detection research was directed toward long-term goals such as improving sensitivity, efficiency, and other capabilities, rather than the short-term goal of matching current capabilities by alternative means.[58]

The neutron detectors used for homeland security present both an opportunity and a challenge. The large base of already deployed equipment, if retrofit with an alternative technology, could be a substantial source of recycled helium-3 for other uses. At the same time, the scale of planned future deployments presents a potentially large future demand for helium-3 if suitable alternatives are not identified. For retrofitting, any alternative would need to match the dimensions, power requirements, and other characteristics of the existing technology. For future deployments, especially beyond the near term, some of these requirements might be relaxed or altered. In either case, the large number of systems means that any alternative would need to be relatively inexpensive. In addition, because of how and where the equipment is used, any alternative would need to be rugged, safe, and reliable. The DNDO plans to deploy alternative neutron detection technology in 2012 to help relieve some of the helium-3 shortage.[59] It has not announced the scale of this deployment or the helium-3 savings that might result.

These requirements for alternative neutron detectors for homeland security raise several policy questions:

- To what performance standard should new detectors be held? If the actual performance of currently deployed systems exceeds the specification required when they were procured, should new detectors be required to match the actual performance or just the specification? If current actual performance in the field does not meet the specification, should new detectors be required to meet the specification anyway?
- To what extent could a reduced performance standard be acceptable to meet other criteria, such as cost, toxicity, and making helium-3 available for other needs?
- How should the performance of the new systems be evaluated? To what extent is acquiring experience in the field sufficient, or is formal testing necessary prior to deployment?
- Should new technology be retrofit to existing systems or used only for new deployments? When considering this question, how should the long-term value of recovered helium-3 be balanced against the short-term costs of retrofitting? What operational reasons exist that would

prefer using a single type of detector rather than a mixture of old and new systems?

The requirements for applications outside the homeland security arena vary and may raise different policy concerns. In the scientific realm, requirements such as large detector areas and rapid data rates may be paramount.[60] Industrial applications may need to withstand vibrations and operate in constrained spaces.

### *Alternative Technologies for Other Applications*

For medical imaging, alternative gases exist that could potentially be used in much the same way as helium-3. The most discussed example is xenon-129, which may be a viable alternative for some medical imaging applications within 5 to 10 years.[61] The required properties of the inhaled gas are very specific, however: polarizability, no toxicity or anesthetic effect, and a high signal-to-noise ratio for the MRI process. Helium-3 is not the only gas with these properties, but it has increasingly become the preferred choice.

For some scientific applications, no satisfactory alternative may exist. For example, helium-3 is the only known way to accomplish very low-temperature cryogenics research.

### *Encouraging Development and Implementation of Alternative Technologies*

Congress could encourage the development and implementation of alternative technologies in a variety of ways. For federal uses such as homeland security, Congress could simply direct agencies to change the technology they use. For the private sector, it could mandate or provide incentives for the use of alternatives. Congress could also attempt to stimulate future availability by funding research and development or demonstration projects or by giving the private sector incentives to do so. Widespread federal use of alternatives might indirectly increase their availability and desirability for the private sector.

## Adjusting the Timing of Demand

In addition to whatever helium-3 is currently in the stockpile, the decay of the existing tritium supply produces new helium-3 each year. If a need for helium-3 can be deferred, future production might meet that need even if the current helium-3 supply cannot. In other words, stretching out deployment schedules might defer helium-3 demand enough that future supplies could meet it. This approach could be an adjunct to the options already described for increasing the future helium-3 supply.

Deferring federal demand is most likely to be a successful approach if a particular helium-3 use is a one-time event or varies significantly from year to year. In such cases, delaying deployment could smooth out the fluctuating demand such that its peaks are below a sustainable level of annual production. One example that meets this description is the continuing deployment of radiation portal monitors by DHS. That deployment is likely to be self-limiting: the number of locations where DHS could usefully deploy the radiation portal monitors is limited, and once all those locations have received a portal, the need for helium-3 to field new portals appears likely to decline (as indicated in Figure 2).

Adjusting deployment schedules is not a short-term solution to the helium-3 shortage, and even in the long term it is unlikely to be a solution unless additional supply is located. For example, even if DHS indefinitely postponed radiation portal monitor deployment, the remaining demand for helium-3 would still exceed the supply. However, adjusting the timing of demand might contribute to a solution that also included other supply and demand measures.

## ALLOCATING THE HELIUM-3 SUPPLY TO USERS

As described above, NNSA extracts helium-3 during the maintenance of the nation's tritium supply and transfers it to the DOE Office of Isotope Production and Research. That office disburses helium-3 to users through two mechanisms: public auctions and interagency transfers. At present, users can receive helium-3 through these mechanisms only if they have an allocation approved by the interagency policy committee (IPC). Congress may wish to consider the criteria that the IPC uses to prioritize these allocations. Alternatively, it could choose to move to a market-based system in which, as with other commodities, the market price of helium-3 establishes the balance between supply and demand and determines the allocation of the limited supply.

## Current Public Auction System

The Office of Isotope Production and Research provides helium-3 to nongovernmental purchasers by public auction. The amount of helium-3 it sells varies from year to year, but, as already noted, increased significantly after 2001, mostly because companies needed more helium-3 to supply neutron detectors to the government for homeland and national security. The auction price for helium-3 has remained relatively steady at less than $100 per liter.[62]

Note that under this model, when DHS, DOE, or another federal agency purchases a device such as a radiation portal monitor, the vendor provides the equipment complete with the helium-3 required to operate it. The vendor purchases that helium-3 from a gas supplier, just as it purchases the other components of the device from other component suppliers. The gas supplier purchases the helium-3 from the federal government. Thus a federal agency that purchases such equipment is, in effect, purchasing helium-3 that was originally sold at auction by DOE.

Although the auction process is public, few companies have historically been interested in purchasing helium-3 and performing any necessary remaining refining steps. Indeed, most of the auction sales have been to just two companies: Spectra Gases Inc., now part of The Linde Group, and GE Reuter Stokes Inc. Many of the auctions have had just one purchaser. Spectra Gases owns the only facility in the United States that is licensed by the Nuclear Regulatory Commission to remove trace amounts of tritium from raw helium-3.

Several factors appear to have influenced the low price of helium-3. Until the shortage was recognized a few years ago, the size of the stockpile seemed large. For most of the history of the program, the supply was sufficient each year to meet the demand, and it was often large enough that the stockpile grew. Projections of future demand, even if they showed increases, may have been unclear or uncertain. Because the production cost of helium-3 was subsidized by the nuclear weapons program, there was no need to recoup the costs of production and extraction. The Office of Isotope Production and Research is responsible for funding its operations through isotope sales.[63] Its need to match its revenue with its expenses may have contributed to its focusing on how to sell helium-3 efficiently, rather than on conserving the helium-3 stockpile or maximizing the stockpile's long-term value to the Treasury, neither of which was its mission.

## Current System for Interagency Distribution

In addition to public auction, helium-3 from the stockpile is sometimes transferred within DOE or to other agencies to meet specific agency needs. For example, in 2008, DOE's Spallation Neutron Source research facility received 35,000 liters of helium-3 for use in scientific neutron detectors. Such transfers are direct and do not use the public auction mechanism. Similarly, in 2009, NNSA and DHS received allocations of helium-3 for homeland and national security needs.

The NNSA and DHS allocations in 2009 took a different approach from the model described above. In the pre-2009 model, when an agency purchased equipment, it was up to the equipment vendor to obtain the helium-3 required to operate the equipment. In the new approach, NNSA and DHS were given helium-3 allocations that limited their equipment purchases. When they contracted with vendors to supply equipment, the contracted vendors could withdraw the allocated amount of helium-3 from the stockpile at a specified price, without going through the auction process.

## Options for Allocation by the Government

One approach to allocating a scarce commodity is to control its use according to national policy priorities. This approach is the one currently taken by the interagency committee, which both determines the priorities and implements the allocation. Other organizational arrangements for prioritization and implementation are also possible, however. Establishing the criteria for prioritization is a key policy decision, because users that fall below a priority cutoff will likely have little or no access to helium-3.

### *Procedural Options for Government Allocation*

At present, the IPC both determines the policy priorities for allocating helium-3 and implements the allocation process. Congress could establish new mechanisms for these tasks. For example, it could establish prioritization criteria directly through legislation, it could direct DOE or another executive branch agency to establish such criteria through the regulatory process, or it could mandate an external advisory committee or a public consultation process. It could assign responsibility for prioritization and implementation to a single agency, or to multiple agencies, rather than to the IPC, which lacks specific statutory authority. Even if Congress determined that the current IPC system is satisfactory, it might wish to formalize it in statute. Alternatively, if Congress concluded that the current system is working well, it might choose to focus its efforts on oversight, rather than new legislation or new programs.

### *Criteria for Government Allocation*

Any allocation system will struggle to balance the complex relationships among multiple competing priorities: science versus security, the private sector versus the public sector, and national needs versus international obligations. Similarly, an allocation system might weigh whether or not an alternative approach exists to helium-3 use.

### Science vs. Security

Science and security are the two largest applications for helium-3. Large scientific facilities designed to use helium-3 may not be able to function without it, and investment in their infrastructure might be wasted if a lack of helium-3 prevented them from operating as planned. For individual scientists whose research requires helium-3, the inability to obtain it might hinder scientific progress and career advancement. On the other hand, security applications such as radiation portal monitors at the U.S. border aim to protect the nation against nuclear and radiological terrorism. It is unclear how to balance these two priorities. The IPC has established an application's contribution to increased national or homeland security as one criterion for receiving a helium-3 allocation. At the same time, however, it considers whether helium-3 is needed to complete infrastructure investments, a criterion that appears to favor several large scientific facilities.

For both science and security, some uses may have diminishing returns. For example, once neutron detectors are deployed at multiple locations on the border, the incremental security benefit of deploying additional detectors may be less. Similarly, if the leading cryogenic scientists are able to obtain helium-3 for their research, providing supplies to newly established laboratories may be less important.

### Public vs. Private

Most private-sector users of helium-3 have designed and developed equipment and procedures based on an expectation that the federal government will make helium-3 available. For example, the oil industry previously used other neutron detection materials for well logging, but has now converted almost entirely to the use of helium-3. Meanwhile, the shortage has resulted largely from increased use of helium-3 by the public sector for radiation detection equipment. Policymakers might decide to prioritize private-sector uses because of their economic importance, because it is mostly the public sector that created the shortage, or because of a general sense that the private sector should come first. Alternatively, they might prioritize public-sector uses because of national needs such as security, because it is the public sector that produces the nation's helium-3 supply, or because of the availability of previously employed alternative methods.

### National vs. International

The United States has domestic needs for helium-3, but it is also involved in international agreements and collaborations that require helium-3. In some cases, such as the International Atomic Energy Agency (IAEA), the purpose of these international uses is security. In others, such as the International Thermonuclear Experimental Reactor (ITER), the aims are scientific. A fundamental question is whether the United States should provide helium-3 to international partners when the domestic supply is restricted. International transfers of helium-3 in support of the IAEA have been scaled back because of the shortage.[64] On the other hand, international security organizations contribute to U.S. security, and international scientific collaborations may provide unique facilities for U.S. scientists.

### Alternatives to Helium-3

Some uses of helium-3 have viable but less optimal alternatives, while others do not. For example, scientists at the DOE national laboratories are assessing alternative detection systems for use in homeland security applications.[65] Scientists attempting to perform ultra-low temperature research have no comparable alternative. An allocation system might weigh the ability to switch to an alternative technology as lessening its priority for helium-3 disbursement. This might elevate unique uses to a higher priority, allowing such work to continue. In contrast, such a criterion might lower the disbursement priority applications deemed by some policymakers as particularly important, for example homeland or national security uses where alternatives exist.

Such a criterion would depend heavily on how policymakers define the acceptable performance and technology availability for an alternative to helium-3.

## Allocation by the Market

A very different approach would be to allow the market price of helium-3 to determine its use. If the demand increased, the price for the limited supply of helium-3 would increase. This might serve to price some users out of being able to purchase helium-3. On the other hand, a high price would allow all users access to the available supply on equal terms and would provide economic incentives to use that supply sparingly and seek out alternatives. A high market price would also provide incentives to develop new sources of supply, although that incentive might not be effective if the production of helium-3 remained a government function.

Congress might identify applications that are essential for policy reasons but that could not afford a sufficient supply of helium-3 at the market price. In that case, it could choose to subsidize the purchase of helium-3 for such applications through direct funding, tax credits, or other measures. While the need for such subsidies might create complexity and increase administrative burdens, it might have the benefit of clarifying the economic costs of policy decisions.

For government users that currently obtain helium-3 through direct transfer rather than purchase, moving to a market model for allocation would increase costs. At the same time, however, the sale of government-produced helium at an increased price would increase revenue. Determining the likely net effect of increased costs and increased revenue would require further analysis and a more detailed picture of how the government would interact with the helium-3 market.

A key component of the helium-3 shortage is that, until recently, the market price for helium-3 did not appreciably fluctuate as demand increased. In part, this is because helium-3 is a byproduct of the U.S. nuclear weapons program, which subsidizes its cost of production. In part, it is because only a few commercial entities have historically been interested in obtaining helium-3 from the government to meet commercial needs. Finally, it may be significant to policymakers that the federal office responsible for sales of helium-3 does not produce the helium-3, is not responsible for maintaining a stockpile of helium-3, and is not in charge of determining the commercial or government-wide need for helium-3.

One challenge for a more market-based system is the interface between the market and the government. On the surface, the public auction process currently in place resembles a market process. In practice, it has several deficiencies. Because the government produces helium-3 as a byproduct of tritium production for the nuclear weapons program, production costs and market demand have little or no influence on the price the government is willing to accept or the amount it is willing to produce. Although the process is structured as an auction, the number of potential buyers is very small. The agency managing the auctions has incentives to match its revenues with its operating costs, rather than to recover the government's production costs or to maximize the revenue to the Treasury in the short or long term. The agency managing the auctions may not have full access to information about supply and demand, even information that already exists in other federal agencies.

It is unclear how effectively a market system would incentivize the private sector to develop new sources of helium-3 or new technologies to replace helium-3. The current market price for helium-3 obtained outside the federal allocation system is already much higher than the historical market price and already reflects the scarcity of supply relative to demand. Nevertheless, there is as yet no sign of significant private-sector involvement in development of new production or utilization technologies. It may be that the market for helium-3 is simply too small to be attractive to the private sector, or it may be that uncertainty about future government decisions is dissuading the private sector from becoming involved until the situation is clearer. For example, private companies might not wish to invest capital in facilities to separate helium-3 from natural gas if they fear that the government will increase its production of tritium through the weapons program, or if they fear that the government may facilitate or subsidize purchases of helium-3 from other countries. To offset such uncertainty, the government might provide incentives to private producers through subsidies, tax credits, or other means.

## SUMMARY OF ISSUES FOR CONGRESS

In summary, Congress could address the helium-3 shortage in a variety of ways:

- Oversight of the current situation, how it arose, and the processes currently in place for addressing it.
- Actions to increase supply:
    - Mandate or encourage increased tritium manufacture by the government or by the private sector through measures such as subsidies or tax incentives.

    - Facilitate access to international sources, perhaps through support for international agreements.
    - Seek alternative sources: commission studies, fund R&D projects or demonstration plants, provide subsidies or incentives.
    - Mandate or facilitate recycling.
- Actions to reduce demand:
    - Fund or encourage the development of alternative technologies.
    - Require or provide incentives for the use of alternative technologies.
- Choices about how the supply should be allocated:
    - Establish a process and criteria for prioritized allocation by the government.
    - Establish a market-based process for allocation.

## ACKNOWLEDGMENTS

The authors wish to acknowledge the valuable assistance of CRS intern Nayanee Gupta.

## End Notes

[1] *CRC Handbook of Chemistry and Physics*, 79th edition (CRC Press, 1998).

[2] It takes about 12.3 years for half of any given quantity of tritium to decay into helium-3. This corresponds to an annual decay rate of about 5.5%. (*CRC Handbook of Chemistry and Physics*)

[3] Well logging is the measurement of geological formations penetrated by a borehole. The use of neutron sources and neutron detectors in well logging can indicate properties such as rock porosity and the presence of hydrocarbons.

[4] "Take a Deep Breath of Nuclear Spin," *CERN Courier*, October 2, 2001. See also, Edwin J. R. van Beek and Jim M. Wild, "Hyperpolarized 3-Helium Magnetic Resonance Imaging to Probe Lung Function," *Proc. Am. Thorac. Soc.*, 2 (6), 528 (2005).

[5] Two Nobel Prizes in Physics have been awarded for work related to such low-temperature physics. In 1996, David M. Lee, Douglas D. Osheroff, and Robert C. Richardson won the Nobel Prize in Physics for their discovery of superfluidity in helium-3. In 2003, Alexei A. Abrikosov, Vitaly L. Ginzburg, and Anthony J. Leggett won the Nobel Prize in Physics for pioneering contributions to the theory of superconductors and superfluids.

[6] House Committee on Science and Technology, Subcommittee on Investigations and Oversight, *The Science of Security Part II: Technical Problems Continue to Hinder Advanced Radiation Monitors*, hearing held November 17, 2009.

[7] William F. Brinkman, Director, DOE Office of Science, and William K. Hagan, Acting Director, DNDO, testimony before the House Committee on Science and Technology, Subcommittee on Investigations and Oversight, April 22, 2010.

[8] Ibid.

[9] William K. Hagan, Acting Deputy Director, DNDO, testimony before the House Committee on Science and Technology, Subcommittee on Investigations and Oversight, November 17, 2009.

[10] Limited supplies of helium-3 are available commercially outside the allocation process through the sale of supplies already held privately, the recycling of helium-3 from defunct equipment, and the purchase of helium-3 from international sources. Some reports indicate a market price for this helium-3 of up to $2,000 per liter, 10 or more times higher than the historical norm. (See, for example, W.P. Halperin, Department of Physics, Northwestern University, testimony before the House Committee on Science and Technology, Subcommittee on Investigations and Oversight, April 22, 2010.)

[11] For example, except for the U.S. mixed oxide fuel facility, all other U.S. international nuclear safeguards support is on hold as a result of the helium-3 supply shortage. William F. Brinkman, Director, DOE Office of Science, testimony before the House Committee on Science and Technology, Subcommittee on Investigations and Oversight, April 22, 2010.

[12] For example, DOE, "Helium 3 Gas Recycling from Neutron Detectors," sources sought notice, February 27, 2009, https://www.fbo.gov/index?id=d9dcf3dacc5b16d99507e0fe01570f0e.

[13] For example, DHS, "He3 Purification, Handling, Tracking and Storage for Neutron Detectors," notice of intent, April 15, 2009, https://www.fbo.gov/index?id=268f2a1f9267724ea41f4f82cd1dec8c.

[14] House Committee on Science and Technology, Subcommittee on Investigations and Oversight, *Caught by Surprise: Causes and Consequences of the Helium-3 Supply Crisis*, hearing held April 22, 2010.

[15] Other nuclear weapon states may possess limited stockpiles of helium-3.

[16] The NNSA, established in 2000, is a semi-autonomous agency in the Department of Energy that is responsible for stewardship of the U.S. nuclear weapons stockpile. See http://nnsa.energy.gov/.

[17] An amount of helium-3 expressed in liters means the amount that would take up that volume in liters if stored at standard temperature and pressure. In practice, helium-3 is typically stored in pressurized tanks that take up a much smaller volume.

[18] For an overview of the debate surrounding the DOE production of tritium in the 1990s, see CRS Report RL30425, *The Department of Energy's Tritium Production Program*, by Richard E. Rowberg (out of print; available from the authors).

[19] See Savannah River Site, Tritium Extraction Facility factsheet, http://www.srs.gov/general/news/factsheets/tef.pdf, December 2007.

[20] DOE FY2011 congressional budget justification, vol. 1, p. 146.

[21] Adrian Cho, "Helium-3 Shortage Could Put Freeze on Low-Temperature Research," *Science*, November 6, 2009, pp. 778-779.

[22] Nuclear Regulatory Commission, personal communication, June 25, 2010.

[23] CRS calculation based on an estimated yield of 0.75 grams of tritium per TPBAR on a typical 18-month cycle.

[24] Nuclear Regulatory Commission, personal communication, June 25, 2010.

[25] CRS calculation as above.

[26] The additional helium-3 production would reach 50% of its long-term level after the increased tritium production had been ongoing for about one tritium half-life, or 12.3 years. It would reach 75% of its long-term level after about two tritium half-lives, or 24.6 years. It would reach 90% of its long-term level after about 41 years. (CRS calculation)

[27] Department of Energy, Office of Inspector General, "Modernization of Tritium Requirements Systems," DOE/IG-0632, December 2003, p. 4.

[28] "World List of Nuclear Power Plants," *Nuclear News*, March 2010.

[29] Dual-use materials have both a military and a civilian use. The United States and other countries often regulate the export of dual-use technologies.

[30] Steve Fetter, Office of Science and Technology Policy, "Overview of Helium-3 Supply and Demand," presentation at the American Association for the Advancement of Science Workshop on Helium-3, April 6, 2010, http://cstsp.aaas.org/files/he3_fetter.pdf.

[31] Richard Kouzes and Edward Siciliano, "Accelerator Based $^3$He Production," appendix in R.L. Kouzes, Pacific Northwest National Laboratory, "The $^3$He Supply Problem," PNNL-18388, April 2009. The example considered in this paper indicates that an accelerator costing several million dollars would take several years to produce even a few tens of liters of helium-3.

[32] Department of the Interior, U.S. Geological Survey, *Mineral Commodity Summaries 2010*, http://minerals.usgs.gov/ minerals/pubs/mcs/2010/mcs2010.pdf, p. 72.

[33] For more information, see Government Accountability Office, *Helium Program: Key Developments Since the Early 1990s and Future Considerations*, GAO-10-700T, May 13, 2010; and National Research Council, Selling the Nation's Helium Reserve, 2010.

[34] *Mineral Commodity Summaries 2010.*

[35] For example, P.C. Hendry and P.V.E. McClintock, "Continuous Flow Apparatus for Preparing Isotopically Pure He-4," *Cryogenics*, vol. 27, pp. 131-138 (1987).

[36] *Mineral Commodity Summaries 2010.*

[37] Thomas A. Davidson and David E. Emerson, Bureau of Mines, Department of the Interior, *Method and Apparatus for Direct Determination of Helium-3 in Natural Gas and Helium*, Report of Investigations 9302, 1990.

[38] CRS calculation. See also Layton J. Wittenberg, Fusion Technology Institute, University of Wisconsin–Madison, "Non-Lunar $^3$He Resources," July 1994, p. 9, which estimates a total reserve of 700 kilograms of helium-3 or approximately 5.7 million liters.

[39] Layton J. Wittenberg, Fusion Technology Institute, University of Wisconsin–Madison, "Non-Lunar $^3$He Resources," July 1994, p. 11, converted from kilowatt-hours (kWh) to dollars at a typical industrial cost for electricity of 7¢/kWh.

[40] Ibid., also at 7¢/kWh.

[41] Richard L. Garwin, "He-3 from Government-Supplied Commercial Helium," paper prepared for the American Association for the Advancement of Science Workshop on Helium-3, April 6, 2010, adjusted by CRS to use an electricity cost of 7¢/kWh for comparability with other estimates.

[42] David Brown, "Lack of Helium-3 Sounding Alarms," AAPG *[American Association of Petroleum Geologists]* Explorer, October 2010.

[43] Richard L. Garwin, "He-3 from Government-Supplied Commercial Helium."

[44] B.M. Olivera, James G. Bradley, and Harry Farrar IV, "Helium Concentration in the Earth's Lower Atmosphere," Geochimica et Cosmochimica Acta, vol. 48, no. 9, September 1984, pp. 1759-1767.

[45] Layton J. Wittenberg, "Non-Lunar $^3$He Resources," p. 12.

[46] Nuclear Regulatory Commission, *Backgrounder on Tritium Production*, http://www.nrc.gov/reading-rm/doc-collections/fact-sheets/tritium.pdf.

[47] Government Accountability Office, *Nuclear Weapons: National Nuclear Security Administration Needs to Ensure Continued Availability of Tritium for the Weapons Stockpile*, GAO-11-100, October 2010.

[48] Savannah River Site, Tritium Extraction Facility factsheet.

[49] For example, Canada has historically prohibited tritium exports to the United States. See "Canada Unlikely to Resolve Any U.S. Tritium Shortage," *The Telegraph*, January 14, 1989.

[50] William F. Brinkman, Director, DOE Office of Science, testimony before the House Committee on Science and Technology, Subcommittee on Investigations and Oversight, April 22, 2010.

[51] Ibid.

[52] Gregory C. Slovik, DNDO, and John Pantaleo, DOE, "Summary of Helium-3's Total Requirements," workshop presentation, March 24, 2009.

[53] Special nuclear material is defined by Title I of the Atomic Energy Act of 1954 (P.L. 83-703) as plutonium, uranium-233, or uranium enriched in the isotopes uranium-233 or uranium-235.

[54] For more information, see CRS Report RL34574, *The Global Nuclear Detection Architecture: Issues for Congress,* by Dana A. Shea; and CRS Report RL31957, *Nonproliferation and Threat Reduction Assistance: U.S. Programs in the Former Soviet Union*, by Amy F. Woolf.

[55] William F. Brinkman, Director, DOE Office of Science, testimony before the House Committee on Science and Technology, Subcommittee on Investigations and Oversight, April 22, 2010.

[56] Well logging is the measurement of geological formations penetrated by a borehole. The use of neutron sources and neutron detectors in well logging can indicate properties such as rock porosity and the presence of hydrocarbons.

[57] Thomas R. Anderson, GE Reuter Stokes, testimony before the House Committee on Science and Technology, Subcommittee on Investigations and Oversight, April 22, 2010.

[58] For more information on the available alternatives, including their advantages, disadvantages, and possible cost, see R.M. Van Ginhoven, R.T. Kouzes, and D.L. Stephens, Pacific Northwest National Laboratory, "Alternative Neutron Detector Technologies for Homeland Security," PIET-43741-TM-840, PNNL-18471, June 9, 2009; and William K. Hagan, Acting Director, DNDO, testimony before the House Committee on Science and Technology, Subcommittee on Investigations and Oversight, April 22, 2010.

[59] Warren M. Stern, Director, DNDO, testimony before the House Committee on Homeland Security, Subcommittee on Emerging Threats, Cybersecurity, and Science and Technology, September 30, 2010.

[60] For example, see Oak Ridge National Laboratory, "PartTec to Market SNS-Developed Neutron Detector System," *EurekAlert* press release, March 11, 2010, http://www.eurekalert.org/pub_releases/2010-03/drnl-ptm031110.php.

[61] Jason C. Woods, Washington University in St. Louis, testimony before the House Committee on Science and Technology, Subcommittee on Investigations and Oversight, April 22, 2010. This source also discusses the limitations of xenon-129 as an alternative to helium-3 for medical imaging applications.

[62] Prices have typically been between \$70 and \$85 per liter. Exact prices are not made public because they are considered business sensitive. (NNSA, personal communication, February 1, 2010)

[63] The Office of Isotope Production and Research sells many different isotopes, not only helium-3, and is the source of many different isotopes in limited supply and demand.

[64] William F. Brinkman, Director, DOE Office of Science, testimony before the House Committee on Science and Technology, Subcommittee on Investigations and Oversight, April 22, 2010.

[65] R.M. Van Ginhoven, R.T. Kouzes, and D.L. Stephens, Pacific Northwest National Laboratory, "Alternative Neutron Detector Technologies for Homeland Security," PIET-43741-TM-840, PNNL-18471, June 9, 2009

In: Helium-3: Critical Uses and Supply Challenges ISBN: 978-1-61470-823-0
Editors: Gabriel A. Ross 

*Chapter 2*

# Managing Critical Isotopes: Weaknesses in DOE's Management of Helium-3 Delayed the Federal Response to a Critical Supply Shortage

***United States Government Accountability Office***

## Why GAO Did This Study

Helium-3 gas is a key component of equipment used at ports and border crossings to detect radiation and prevent the smuggling of nuclear material into the United States, among other uses. The National Nuclear Security Administration (NNSA), a separate agency within the Department of Energy (DOE), extracts helium-3 and controls the inventory. Since 2003, NNSA has made helium-3 available for sale to DOE's Isotope Development and Production for Research and Applications Program (Isotope Program). After September 11, 2001, demand increased for radiation detection equipment, and in 2008, the federal government learned that it faced a severe domestic shortage of the gas. GAO was asked to review DOE's management of helium-3 to (1) determine the extent to which the federal government's response to the helium-3 shortage was affected by DOE's management of helium-3; (2) determine the federal government's priorities for allocating the limited supply of helium-3; and (3) describe the steps that the federal government is taking to increase the helium-3 supply and develop alternatives to helium-3. GAO reviewed DOE and NNSA documents and interviewed cognizant agency officials.

## WHAT GAO RECOMMENDS

GAO recommends, among other things, that DOE clarify whether the stewardship for those isotopes produced outside the Isotope Program, such as helium-3, rests with the program or another DOE entity. DOE stated that it understands and can implement these recommendations.

## WHAT GAO FOUND

The federal government's awareness of and response to the helium-3 shortage was delayed because no DOE entity had stewardship responsibility for the overall management of helium-3—a by-product of the radioactive decay of tritium, a key component of the U.S. nuclear weapons program. Although the Isotope Program's mission includes selling isotopes and providing related isotope services, senior program officials said that they interpret this mission to exclude helium-3 and 16 other isotopes that the program sells but whose supply it does not control. As a result of this weakness in DOE's management of helium-3, officials at the Isotope Program and NNSA did not communicate about the helium-3 inventory or its extraction rate. According to NNSA and Isotope Program officials, they communicated with each other about how much helium-3 to sell each year and at what price but not about the size of the helium-3 inventory or extraction rate because NNSA generally treated this information as classified, due to concerns that the helium-3 inventory could be used to calculate the size of the U.S. tritium stockpile. NNSA and Isotope Program officials told GAO that this lack of communication contributed to the federal government's delayed response to the helium-3 shortage. The standards for internal control in the federal government state that information should be communicated to management and others within a time frame that enables them to carry out their responsibilities. Further, without stewardship by a DOE entity, key risks to managing helium-3, such as the lack of complete information on the production and inventory of helium-3, were not identified or mitigated. The federal standards for internal control state that management should assess the risks faced from external and internal sources and decide what actions to take to mitigate them.

Facing this critical shortage of helium-3, DOE and other federal agencies are collaborating to bring supply and demand into balance. Specifically, in July 2009, an interagency policy committee was formed, which halted allocations of helium-3 for domestic radiation detection equipment and established three priorities for allocating helium-3: (1) applications for which there are no alternatives to helium-3 have first priority (e.g., research that can be achieved only with helium-3); (2) programs for detecting nuclear material at foreign ports and borders have second priority; and (3) programs for which substantial costs have already been incurred have third priority (e.g., a DOE research facility that conducts physics research).

To increase the supply of helium-3, the federal government is, among other things, pursuing other sources and developing alternatives. Specifically, NNSA is in discussions with Ontario Power Generation (OPG), a power company in Ontario, Canada, to obtain helium-3 from its stores of tritium. OPG has accumulated tritium as a by-product of producing electricity using a type of nuclear reactor not found in the United States. Also, federal agencies and private companies are researching alternative technologies to replace helium-3 in several applications to decrease demand.

## ABBREVIATIONS

| | |
|---|---|
| cGMP | current good manufacturing practices |
| DHS | Department of Homeland Security |
| DOD | Department of Defense |
| DOE | Department of Energy |
| Isotope Program | Isotope Development and Production for Research and Applications Program |
| MOU | Memorandum of Understanding |
| NIH | National Institutes of Health |
| NIST | National Institute of Standards and Technology |
| NNSA | National Nuclear Security Administration |
| OPG | Ontario Power Generation |

May 12, 2011

The Honorable Brad Miller
Ranking Member
Subcommittee on Energy and Environment
Committee on Science, Space, and Technology
House of Representatives

The Honorable Donna Edwards
Ranking Member
Subcommittee on Investigations and Oversight
Committee on Science, Space, and Technology
House of Representatives

Helium-3 gas is a critical component of radiation detection equipment, including radiation detection portal monitors that are used to screen cargo and vehicles at ports and border crossings around the world to prevent nuclear material from being smuggled into the United States. For example, the Department of Homeland Security (DHS) has deployed over 1,400 radiation detection portal monitors containing helium-3 at U.S. ports and border crossings. In addition, helium-3 is used in various industrial applications, such as oil and gas exploration and road construction, and in research applications, including physics research requiring ultra-low temperatures that can only be achieved using helium-3. Until 3 years ago, the United States' helium-3 supply was viewed as sufficient to meet demand. Outside the United States, Russia has been the only other major source of commercially available helium-3. But in 2008, the federal government abruptly learned that it faced a severe shortage of helium-3. At the same time Russia has curtailed its sales of helium-3, indicating that its supply is also waning. Because of this global shortage, no future deployments of radiation detection portal monitors containing helium-3 are planned in the United States.

Helium-3 is a byproduct of the radioactive decay of tritium,[1] a key component of the nation's nuclear weapons that is used to enhance their power. The U.S. tritium stockpile is maintained by the National Nuclear Security Administration (NNSA), a separately organized agency within the Department of Energy (DOE).[2] In maintaining this tritium stockpile, NNSA removes the helium-3 that accumulates as tritium decays because the helium-3 can diminish the effectiveness of the nuclear weapons. In the past, NNSA and its predecessor agencies considered helium-3 to be a waste product of the weapons program and vented it to the atmosphere. From about 1980 through 1995, and again from 2003 through 2008,[3] NNSA provided helium-3 to DOE's Isotope Development and Production for Research and Applications Program (Isotope Program) to sell. The Isotope Program's mission is to produce and sell isotopes and related isotope services, maintain the infrastructure required to do so, and conduct research and development on new and improved isotope production and processing techniques.[4] DOE's Isotope Program produces and sells about 200 isotopes, though it does not control the supply—that is, the production or inventory—of all the isotopes it sells, such as helium-3, which is extracted from tritium by NNSA.

With the end of the Cold War, the United States has been reducing the number of nuclear weapons in its stockpile, resulting in less tritium and, therefore, less helium-3. In the aftermath of the September 11, 2001, terrorist attacks, however, demand for helium-3 increased due to the deployment of radiation detection portal monitors at ports and border crossings in the United States and other countries to prevent the smuggling of nuclear material. Thousands of such portal monitors were deployed across the United States and overseas, and more are planned to be deployed. Additionally, large quantities of helium-3 have been used in the last 10 years by research facilities in the United States and internationally for large-scale physics research applications, and more is needed for this research to continue. Overall, this decreasing supply during a period of increasing demand resulted in a shortage of helium-3.

In June 2008, a contractor alerted DHS that the contractor was unable to obtain a sufficient amount of helium-3 to fulfill its contract to provide radiation detection portal monitors. DHS contacted the Isotope Program to inquire about the helium-3 supply, and in September 2008, DOE responded to this sudden awareness of the shortage by suspending its helium-3 sales through public auction, which was the vehicle through which the Isotope Program sold helium-3. In March 2009, DHS, the Department of Defense (DOD), DOE, and NNSA formed an integrated project team to collect and analyze information and make recommendations to senior management at these departments on actions to be taken to address the shortage. The overall shortage of helium-3 raises concerns about the United States' ability to provide a sustainable supply of helium-3 for national security and research applications. In this context, you asked us to review DOE's management of helium-3. Our objectives were to (1) determine the extent to which the federal government's response to the helium-3 shortage was affected by DOE's management of helium-3; (2) determine the federal government's priorities for allocating the limited supply of helium-3 to various users; and (3) describe the steps that the federal government is taking, if any, to increase the helium-3 supply and develop alternatives to helium-3.

## Scope and Methodology

To determine the extent to which the federal government's response to the helium-3 shortage was affected by DOE's management of helium-3, we reviewed the DOE Isotope Program's strategic planning documents, helium-3 sales data, and information on NNSA's inventory of helium-3. We also interviewed officials at DOE, DOE's Savannah River Site, NNSA, and DHS, as well as representatives from Linde[5] and GE Reuter-Stokes, the two principal companies that purchased helium-3 from the Isotope Program. Also, we used the federal standards for internal control to assess DOE's management of helium-3.[6] To determine the federal government's priorities for allocating the limited supply of helium-3 to various users, we reviewed documents of the integrated project team. We also reviewed the helium-3 allocation decisions, criteria, and process of the interagency policy committee convened in 2009 by the National Security Staff, which report to the National Security Advisor,[7] to oversee the integrated project team and make policy decisions to manage the helium-3 shortage. The policy committee is a multi-agency committee consisting of key agencies and departments that use helium-3 applications to support their missions, including the Department of Commerce's National Institute of Standards and Technology (NIST), DOD's Defense Threat Reduction Agency, the Department of State, DOE, and DHS. We also interviewed officials at NIST, the Department of Health and Human Services' National Institutes of Health (NIH), DHS, DOD, DOE, DOE's Oak Ridge and Pacific Northwest National Laboratories, and NNSA, as well as National Security Staff. To describe the steps that the federal government is taking, if any, to increase the helium-3 supply and develop alternatives to helium-3, we reviewed feasibility studies that presented options for alternative sources and recycling unused equipment and interviewed representatives from Ontario Power Generation, a Canadian power company. We reviewed research and documentation, including test results, on alternatives to helium-3 that are being developed by companies and interviewed representatives from these companies and officials at DHS, DOD, DOE, NIH, NIST, NNSA, and Oak Ridge National and Pacific Northwest National Laboratories. We conducted this performance audit from April 2010 to May 2011, in accordance with generally accepted government auditing standards. Those standards require that we plan and perform the audit to obtain sufficient, appropriate evidence to provide a reasonable basis for our findings and conclusions based on our audit objectives. We believe that the evidence obtained provides a reasonable basis for our findings and conclusions based on our audit objectives.

## BACKGROUND

The unique physical properties of helium-3 have led to its use in a wide variety of national security, scientific, industrial, and medical applications. Helium-3 is widely used for detecting nuclear material and safeguarding nuclear weapons because, among other things, of its ability to efficiently absorb neutrons. In radiation detection equipment, helium-3 is used to detect neutrons that are emitted by nuclear material. In radiation detection portal monitors, long, thin metal tubes are filled with helium-3; neutrons passing through these tubes react with the helium-3, creating charged particles that are detected by the monitors. Also, as a nontoxic gas that is not absorbed by the human body, helium-3 is used in magnetic resonance imaging (MRI) to research pulmonary disorders, such as chronic obstructive pulmonary disease.[8]

Helium-3 is rare because it is currently extracted solely as a byproduct of the radioactive decay of tritium. During the Cold War, the United States produced tritium in nuclear reactors and stockpiled it for the nuclear weapons program. Helium-3 is also available from natural sources, such as subterranean natural gas deposits, but it has not been pursued commercially in the past because it is found in very low concentrations. NNSA and its predecessor agencies produced tritium at the Savannah River Site K-reactor in South Carolina and purified tritium by removing helium-3 at the Mound Plant, a weapons research laboratory in Ohio. Following the end of the Cold War, as the United States reduced its nuclear weapons stockpile and ceased producing tritium, its inventory of helium-3 decreased commensurately. In 1988, DOE shut down the K-reactor for safety reasons and, in 1995, closed the Mound Plant, thus eliminating the U.S. government's large-scale ability to produce and purify tritium. NNSA has been able to meet the tritium needs of the nuclear weapons program by maintaining the existing stockpile and recycling tritium from dismantled nuclear warheads.

To maintain the current tritium stockpile, NNSA extracts helium-3 from tritium on a daily basis and stores the helium-3 in pressurized cylinders at the Savannah River Site. To remove trace amounts of tritium and other impurities,

NNSA ships these cylinders of helium-3 to Linde in New Jersey, which operates the only commercial facility in the United States licensed by the Nuclear Regulatory Commission to purify helium-3 of trace amounts of tritium. While NNSA's helium-3 inventory at the Savannah River Site is constantly changing, as of early February 2011, it was almost 31,000 liters. NNSA estimates that about 8,000 to 10,000 liters of helium-3 will be made available per year from the current tritium stockpile. Like the United States, Russia extracts helium-3 from its tritium stockpile. According to National Security Staff documentation and representatives of Linde, Russia has curtailed its sales of helium-3, indicating that its supply is likely waning.

Isotope production and distribution has been a long-standing mission of DOE. DOE's Isotope Program provides isotopes to support the national need for a reliable supply of isotopes used in medicine, industry, and research. DOE transferred the Isotope Program to DOE's Office of Science from DOE's Office of Nuclear Energy starting in fiscal year 2009. In anticipation of this transfer, in August 2008, the Isotope Program organized a workshop to discuss the nation's needs for isotopes, and identified those isotopes with supply challenges. As noted in the workshop summary report, the workshop assembled, for the first time, stakeholders from all the different areas of the diverse isotope community to discuss the nation's current and future needs for isotopes and to consider options for improving the availability of needed isotopes.[9] The workshop enabled the Isotope Program to discuss and develop program priorities, including those isotopes, such as helium-3, that were in short supply. This workshop identified 12 isotopes, including helium-3, that faced supply challenges and had less than 3 years before their supplies at that time were completely consumed.

In August 2003, the Isotope Program signed a Memorandum of Understanding (MOU) with NNSA to make available an initial 103,570 liters of helium-3 for sale, followed by at least 10,000 liters of helium-3 per year from 2004 through 2008. At the time the MOU was signed, NNSA's inventory of helium-3 was estimated at about 260,000 liters. Following this agreement, the Isotope Program held a series of public auctions to sell helium-3. A public auction process was used, according to program officials, to encourage competition. The Isotope Program and NNSA determined the quantity and minimum price of helium-3 for each auction; the price was set to recover the costs to extract helium-3 and the administrative costs of selling it. Linde and GE Reuter-Stokes, a company that manufactures helium-3 tubes for radiation detection portal monitors and other neutron detection applications, have been the two buyers of helium-3 who participated in the Isotope Program's public auctions. From 2003 through 2009, the Isotope Program sold, or NNSA transferred, more than 209,000 liters of helium-3—an average of almost 30,000 liters of helium-3 per year, as shown in table 1. Given NNSA's capacity to extract between 8,000 and 10,000 liters of helium-3 annually, this rate of sale exceeded the extraction rate and decreased the inventory.

Following the terrorist attacks of September 11, 2001, the demand for helium-3 nearly tripled, because of the increased focus on radiation detection applications. Specifically, DHS's Radiation Portal Monitor program, NNSA's Second Line of Defense program, and DOD's Guardian program all use helium-3 in radiation detection portal monitors deployed at domestic and foreign ports, border crossings, and military installations. DHS alone has deployed over 1,400 radiation detection portal monitors domestically. The largest demand for helium-3 has historically been for homeland security and scientific research, but demand for other applications, such as in MRIs for lung research, has also increased.

**Table 1. Quantities of Helium-3 Sold or Transferred from 2003 through 2009**

| Date sold | Customer | Quantity (liters) |
|---|---|---|
| March 2003 | Spectra Gases[a] | 7,826 |
| February 2004 | G.E. Reuter-Stokes | 95,791 |
| June 2005 | Spectra Gases[a] | 23,310 |
| July 2006 | Spectra Gases[a] | 26,963 |
| August 2008 | DOE Spallation Neutron Source (interagency transfer from NNSA) | 34,325 |
| January 2009 | Lind[e] | 10,585 |
| | G.E. Reuter-Stokes | 11,088 |
| **Total** | | **209,888** |

Source: GAO analysis of NNSA and Isotope Program data.

[a]Spectra Gases was acquired by Linde in 2006 and changed its name to Linde in 2009.

# FEDERAL RESPONSE TO THE HELIUM-3 SHORTAGE WAS DELAYED BY WEAKNESSES IN DOE'S MANAGEMENT OF HELIUM-3

The federal government's awareness of and response to the helium-3 shortage was delayed because no DOE entity had stewardship responsibility for the overall management of helium-3. As a result of this lack of stewardship responsibility, officials from DOE's Isotope Program, which sold helium-3, and NNSA, which extracted it from tritium, did not communicate about the helium-3 inventory or its extraction rate. Without stewardship responsibility, key risks to managing helium-3, such as the lack of understanding of the helium-3 inventory and the demand for helium-3, were not identified or mitigated by either entity.

## Officials from DOE's Isotope Program and NNSA Did Not Communicate about the Helium-3 Inventory or its Extraction Rate

While the Isotope Program's mission includes selling isotopes and providing related isotope services, senior program officials said that they interpret this mission to exclude helium-3 and other isotopes that the program sells but whose supply it does not control. Accordingly, Isotope Program officials noted that the program sold and distributed helium-3 solely as a courtesy to NNSA, not because it was a core part of the program's mission or because it believed it had a stewardship responsibility to do so. NNSA officials also noted that helium-3 stewardship was not part of NNSA's mission of managing the nation's nuclear weapons. Without such stewardship responsibility, NNSA and Isotope Program officials did not communicate about the helium-3 inventory or its rate of extraction.

The Isotope Program's management of helium-3 sales was hampered by this lack of communication regarding the size of the helium-3 inventory and the rate at which helium-3 is extracted from tritium. Prior to selling helium-3 at public auction, officials from the Isotope Program and NNSA communicated with each other regarding how much helium-3 would be available to sell that year and the minimum price for which it would be sold at auction. However, Isotope Program and NNSA officials did not discuss the size of the helium-3 inventory, how much was being added to the inventory each year, or how quickly the inventory was being depleted. Additionally, NNSA officials did not inform Isotope Program officials when they transferred more than 34,000 liters of helium-3 in August 2008 to DOE's Spallation Neutron Source, a physics research facility at the Oak Ridge National Laboratory that uses helium-3 in large-scale neutron detectors. This transfer—more than what the Isotope Program sold per year, on average from 2003 to 2009—greatly reduced the helium-3 inventory, but NNSA officials did not inform Isotope Program officials about it until after the transfer was completed. Despite the helium-3 inventory being greatly reduced, in September 2008 the Isotope Program and NNSA renewed their MOU to continue selling helium-3 for an additional 5 years without discussing the size of the helium-3 inventory or the rate at which sales and large transfers—such as the one to DOE's Spallation Neutron Source—were reducing the inventory.

Helium-3 inventory and production information was not shared between officials at the Isotope Program and NNSA because, according to NNSA officials, this information was generally treated as classified by NNSA out of concern that the inventory and annual extraction rate could be used to calculate the size of the U.S. tritium stockpile, which is classified. In describing the situation, Isotope Program officials stated that they did not have the requisite "need to know" to gain access to this information, and consequently, did not discuss it. In other words, Isotope Program officials did not believe that they needed complete information on the size of the helium-3 inventory or how much was being added to the inventory each year in order to carry out the program's mission because helium-3 does not fall within its mission. One of the standards for internal control in the federal government—information and communications—states that information should be recorded and communicated to management and others within an entity in a form and within a time frame that enables them to carry out their responsibilities.[10] The lack of communication between NNSA and Isotope Program officials was not consistent with this standard. Isotope Program and NNSA officials told us that this lack of communication contributed to the government's delayed response to the helium-3 shortage. NNSA officials acknowledged the ambiguity about what information can be communicated about the helium-3 inventory and, in January 2010, issued a memorandum to clarify and broaden what information about helium-3 can be shared publicly. As a result, NNSA now reports that 8,000 to 10,000 liters of helium-3 will be made available per year.

## Without Stewardship Responsibility, Key Risks to Managing Helium-3 Were Not Identified or Mitigated

Senior Isotope Program officials said that they did not identify and mitigate key risks to managing helium-3 sales because, unlike most isotopes that the program sells, officials do not consider stewardship of helium-3 to be part of the program's mission. Specifically, these Isotope Program officials did not consider their lack of understanding of the helium-3 inventory or the demand for helium-3 as risks to managing helium-3 sales. According to these officials, for those isotopes that are included in its mission—the isotopes that it produces—the Isotope Program developed strategic planning documents and generally updated these documents on an annual basis. We reviewed the Isotope Program's strategic planning documents and found examples where the program assessed risks to isotopes that it produces and sells. The Isotope Program, however, did not perform strategic planning for helium-3, including assessing risks, because program officials do not consider stewardship of helium-3 to be part of the program's mission. For those isotopes that the Isotope Program sells but whose supply it does not control, such as helium-3, Isotope Program officials told us that they see their role as a conduit that sells these isotopes to customers as a courtesy. There are 17 isotopes, including helium-3, that the program sells but, according to program officials, does not have stewardship responsibility for because the Isotope Program does not control their supply. For example, lithium-6, which is used in neutron detection applications and battery research, is sold by the Isotope Program, but its supply is controlled by NNSA.

Although program officials do not consider helium-3 to be part of the program's mission, it nonetheless collected information in order to forecast demand. To do this, the Isotope Program recorded the number of telephone inquiries from potential customers and the volume of helium-3 discussed. Because it does not view helium-3 as a part of its mission, program officials said that the Isotope Program did not take proactive steps to solicit information to better understand future demand. For example, according to representatives, the Isotope Program did not solicit information from Linde, the company that had purchased or purified nearly all the helium-3 in the United States, which gave Linde a more complete understanding of the historical and future demand for helium-3. In tracking telephone inquiries, the Isotope Program's records show that it received nine telephone calls in 2008 from customers who were interested in acquiring 1,226 liters of helium-3. In contrast, according to Linde documentation, based on its actual use, the demand for helium-3 in 2008 was nearly 60,000 liters. Linde representatives also noted that the Isotope Program did not seek demand-related information from Linde until after the shortage was realized.

Isotope Program officials told us that forecasting the demand for isotopes is very difficult because demand for isotopes can quickly change. Because of its effect on isotope inventories, changing demand is a risk to the program's management of the sale of all isotopes. One of the federal standards for internal control—risk assessment—states that management should assess the risks faced entity-wide, and at the activity level, from both external and internal sources, and that once risks have been identified, management should decide what actions should be taken to mitigate them. Risk identification methods may include, among other things, forecasting and strategic planning, and consideration of findings from audits and other assessments. A DOE advisory committee has also noted the importance of understanding demand for isotopes. In August 2008, DOE tasked its Nuclear Science Advisory Committee with establishing a standing subcommittee to research the needs and challenges of the Isotope Program and make recommendations to address them. According to the subcommittee report, for the Isotope Program to be efficient and effective, it is essential that it accurately forecasts the demand for isotopes.[11] The report noted that the "ability of the program to predict demand for certain isotopes needs vast improvement." It went on to recommend that the program "maintain a continuous dialogue with all interested federal agencies and commercial isotope customers to forecast and match realistic isotope demand and achievable production capabilities."

Isotope Program officials told us that they are considering convening a workshop, possibly in the summer of 2011, with federal agency stakeholders to discuss supply and demand of all isotopes that are produced or sold by the Isotope Program. According to the Director of the Facilities and Project Management Division, which manages the Isotope Program, while the program does not consider this role a part of its mission, it is volunteering to convene this workshop to be helpful to the isotope user community.

## THE FEDERAL GOVERNMENT'S THREE PRIORITIES FOR ALLOCATING THE LIMITED SUPPLY OF HELIUM-3 EXCLUDE DOMESTIC RADIATION DETECTION PORTAL MONITORS

In July 2009, the National Security Staff, under the National Security Advisor, established an interagency policy committee consisting of officials from DOD, DOE, DHS, the Department of Commerce, and the Department of State to address the helium-3 shortage. In doing so, the policy committee established the following three priorities for allocating the limited supply of helium-3:

- Priority 1: Applications for which there are no alternatives to helium-3, which includes, for example, research that requires ultra-low temperatures that can be achieved only with helium-3.
- Priority 2: Programs for detecting nuclear material at foreign ports and borders, which includes, for example, NNSA's Second Line of Defense program that deploys radiation detection portal monitors at key overseas ports and border crossings.
- Priority 3: Programs for which substantial costs have already been incurred, such as DOE's Spallation Neutron Source research facility that conducts physics research.

Furthermore, the committee eliminated further allocations of helium-3 for domestic radiation detection portal monitors beginning in fiscal year 2010 because, according to committee documents, it determined there are alternatives to using helium-3 to detect neutrons in these portal monitors. The policy committee also determined that it will not support allocating helium-3 for any new applications that would increase the demand for helium-3.

Following this approach, the policy committee has allocated helium-3 to federal agency and commercial customers from 2009 through 2011, as shown in table 2. These allocations have brought the supply and demand of helium-3 into closer balance and mark a significant decrease from the amount the Isotope Program previously sold or transferred from 2003 through 2009—an average of about 30,000 liters per year.

The policy committee developed a process for customers to request allocations of helium-3 using "champions," which are agency officials who represent the federal agencies for which they work and its grantees; a champion is appointed by the Isotope Program for nonfederal customers, such as the oil and gas industry. The champion for a specific category of customer gathers all the helium-3 requests and determines whether the requests are consistent with the policy for allocating helium-3. If so, the champion submits the requests to the policy committee. The policy committee weighs the requests against the helium-3 priorities and the amount of helium-3 that is available to make allocation decisions. After allocation decisions are made, customers are notified and, if they received an allotment, they must submit a request for the helium-3 to Linde, which is contracted by the Isotope Program to purify and distribute the helium-3 allocations. When Linde receives a request for helium-3, according to Linde representatives, they verify the customer's allocation with the appropriate champion. Once verification is received, Linde ships the allotted amount of helium-3 to the customer. This allocation and receipt process is the same for all customers—federal agencies, researchers, and private companies, as shown in figure 1.

The helium-3 champions or the policy committee may reject a request for helium-3, as shown in figure 1. If rejected, customers may resubmit a request the following year. According to seven of the agency champions, customers are aware of the policy committee's priorities and have usually reduced the amount of their request to the absolute minimum amount of helium-3 that is needed. As a result, the committee has approved most requests.

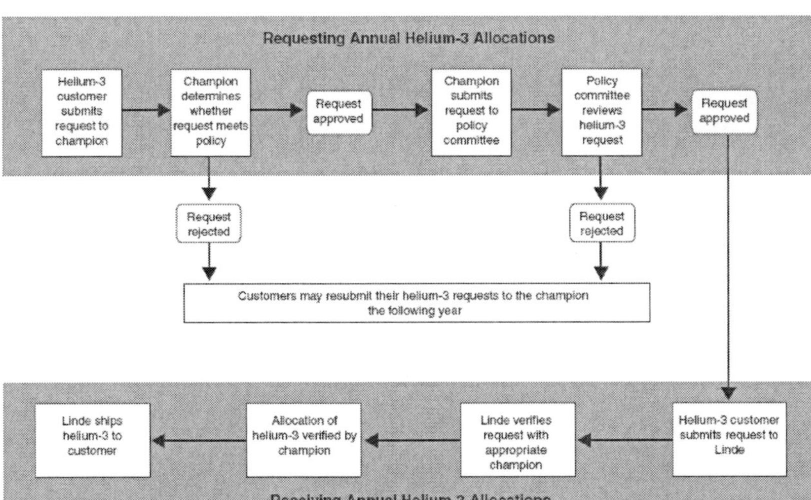


Source: GAO analysis of information from DOE and Linde.

Figure 1. Helium-3 Request and Receipt Process.

**Table 2. Quantities of Helium-3 Allocated and Used, in Liters, from Fiscal Year 2009 to Fiscal Year 2011**

| Customer | Quantities | | |
|---|---|---|---|
| | FY 2009 | FY 2010 | FY 2011 |
| Low temperature research | N/A[a] | 452[b] | 700 |
| DHS | N/A[a] | 438[c] | 1,218 |
| DOD | N/A[a] | 1,530 | 3,521 |
| NNSA | 6,367 | 5,098[d] | 5,791 |
| DOE's Office of Science | 2,400 | 341 | 315 |
| Intelligence Community | N/A[a] | N/A[a] | 763 |
| NIH | N/A[a] | 260[e] | 1,400 |
| NIST | N/A[a] | 607[f] | 236 |
| Oil and gas industry | N/A[a] | 695[g] | 1,000 |
| Road construction industry | N/A[a] | N/A[a] | 350 |
| Total | 8,767 | 9,421 | 15,294 |

Source: GAO analysis of information from the interagency policy committee and Linde.

[a]N/A refers to instances when a customer either did not receive an allocation or received its allocation through another customer.

[b]Low temperature research was allocated 1,800 liters but used only 452 liters.

[c]DHS was allocated 772 liters but used only 438 liters.

[d]NNSA was allocated 5,150 liters but used only 5,098 liters.

[e]NIH was allocated 1,800 liters but used only 260 liters.

[f]NIST was allocated 832 liters but used only 607 liters.

[g]The oil and gas industry was allocated 1,000 liters but used only 695 liters.

**Table 3. Allocated Helium-3 Prices per Liter**

| Customers | 2009 | 2010 | 2011 |
|---|---|---|---|
| Federal agencies and their grantees | $450 | $365 | $600 |
| Commercial and nonfederal agencies | $450 | $365 | $1,000 |
| cGMP | $600 | $485 | $720 |

Source: GAO analysis of information from DOE and Linde.

When determining the annual allocations for helium-3, the policy committee also recommends to the Isotope Program a price at which helium-3 should be sold to different customers—including the federal, medical, and commercial entities. When the policy committee first began to allocate helium-3 in 2009, the Isotope Program, in consultation with the policy committee, established two different prices—one for medical applications and the other for all other applications. For use in the manufacture of drugs, medical devices, and other products, the Food and Drug Administration requires that helium-3 must be certified to meet specific requirements, called current good manufacturing practices (cGMP).[12] In 2009, the price for cGMP helium-3 was $600 per liter; helium-3 for all other applications was priced at $450 per liter.[13] Certifying helium-3 under cGMP requirements is more expensive, according to Linde representatives, because of the extra certification and purity testing that is required. This practice was continued for allocations in 2010. In 2011, however, the Isotope Program divided the price for helium-3 to be used for non-cGMP applications into two categories: one for federal agencies and their grantees and one for commercial and nonfederal agencies. Table 3 shows the different prices for helium-3, per liter, beginning with 2009.

## Key Federal Agencies Are Collaborating to Increase the Helium-3 Supply and Develop Alternatives

DOE and NNSA are taking actions to increase the supply of helium-3 by, among other things, pursuing other sources and recycling helium-3 from retired equipment. Specifically, NNSA officials said that NNSA is in discussions with Ontario Power Generation (OPG) to determine the feasibility of obtaining helium-3 from OPG's stores of tritium. OPG has accumulated this tritium as a by-product of producing electricity using heavy-water nuclear reactors.[14] According to OPG officials, it owns 16 heavy-water nuclear reactors that are currently operating that have produced enough tritium to initially yield approximately 100,000 liters of helium-3. According to NNSA officials, once this initial amount is recovered, OPG estimates that its stores of tritium may yield about 10,000 liters of helium-3 annually. Combined with NNSA's current annual production of helium-3, obtaining helium-3 from OPG could boost the United States' supply to about 18,000 to 20,000 liters per year. In addition, DHS and DOE have expressed interest in exploring the option of extracting helium-3 from natural helium, or helium-4. Helium-3 is found in small quantities in natural helium and could possibly be extracted from the nation's helium reserve near Amarillo, Texas, that is managed by the Department of the Interior's Bureau of Land Management. The Bureau of Land Management estimates that approximately 125,000 liters of helium-3 may be present in the helium reserve, which could be extracted over the next 10 years. DOE officials estimate that a similar reserve of natural helium in Wyoming could yield another 200,000 liters of helium-3 that could be extracted over the life of the reserve. DHS officials note, however, that a feasibility study is needed to determine whether it would be cost-effective to extract helium-3 from natural helium.

Federal agencies and private sector companies have started to recycle helium-3 from unused equipment in order to boost the supply. For example, an analysis conducted by DOE shows that it can extract helium-3 from retired tritium storage beds at its tritium extraction facility at the Savannah River Site. DOE estimates that it could extract 8,000 to 10,000 liters from these storage beds every 8 to 10 years, beginning as early as 2012. Additionally, DOE surveyed its national laboratories and identified over 1,500 liters of helium-3 in unused equipment and storage cylinders that could be reused immediately. DHS has also identified retired equipment from which helium-3 can be extracted. Private companies have also started to recycle helium-3 from decommissioned radiation detection portal monitors. For example, according to a representative from a helium-3 tube manufacturing company, the company is buying retired radiation detection equipment to extract the helium-3.

In addition to increasing the supply of helium-3, federal agencies and private companies are researching alternatives to helium-3 for several applications in order to decrease demand. For example, the government is conducting research to develop alternatives for neutron detection applications, including radiation detection portal monitors and nuclear physics research, which together use more helium-3 than any other application. DHS, DOE, DOD, and NIST, for example, are supporting approximately 30 different programs, some of which may result in technologies available for use in radiation detectors that, according to agency documents, could be ready by 2012.[15] Similarly, DOE's Spallation Neutron Source research facility is coordinating with similar facilities internationally—including those in Germany, Japan, Russia, and Sweden—to develop alternative technologies for large-scale physics research applications. The private sector is also researching alternatives to helium-3 for radiation detection portal monitors and other applications, including MRIs for the lungs. For example, equipment using the isotopes lithium-6 and boron-10 may be able to replace helium-3 in radiation detection portal monitors, according to representatives from companies that are developing them. Like helium-3, lithium-6 is produced by NNSA and sold by the Isotope Program. According to one program official, this official contacted NNSA to inquire about the inventory and production rate of lithium-6 because of the potential for increased demand if lithium-6-related technologies are chosen to replace helium-3 in radiation detection portal monitors. NNSA officials told the Isotope Program official, however, that such information is classified and cannot be shared, but assured the official that NNSA has enough lithium-6 to meet any future increase in demand. In March 2011, the Director of the Facilities and Project Management Division, which manages the Isotope Program, said that, although classified, program officials do have access to this information through NNSA's Office of Nuclear Materials Integration. DOE officials said that they have evaluated the potential demand for lithium-6 and have taken steps to ensure there is an adequate supply. According to its director, the Office of Nuclear Materials Integration is responsible for, among other things, coordinating management of certain isotopes produced by NNSA, including tracking the inventory of these isotopes, and coordinating communication within DOE and NNSA. The director stated that it is the responsibility of the Isotope Program, however, to manage the activities under its control for these isotopes, such as selling them outside DOE and NNSA and conducting 5-year supply and demand forecasts. This raises concerns that without stewardship responsibility for the overall management of the supply and demand of lithium-6, or the other isotopes produced by NNSA, neither the Isotope Program nor any other DOE or NNSA entity may detect an imbalance, resulting in the shortage of another isotope.

## CONCLUSIONS

Facing a critical shortage of helium-3 since 2008, DOE and other federal agencies are collaborating to bring supply and demand into balance, while supporting essential applications for which there are no alternatives. This shortage occurred because the demand for helium-3 rose sharply in response to the increased deployment of radiation detection portal monitors, in addition to the increased use of helium-3 in research and other applications. The amount of helium-3 sold by the Isotope Program quickly outpaced the annual production, and this imbalance went undetected until the supply of helium-3 reached a critical shortage. The overall federal awareness of and response to the helium-3 shortage was delayed because no entity within DOE had stewardship responsibility for coordinating the production and sale of helium-3. Furthermore, there was a lack of communication between NNSA and Isotope Program officials over the size of the helium-3 inventory, how much was added to the inventory annually, and how quickly the Isotope Program's sales were depleting the inventory. While the Isotope Program's mission is to manage the production and sale of isotopes, including developing strategic plans and assessing risks for these isotopes, it has not taken a similar stewardship role in managing the 17 isotopes, including helium-3, whose supply it does not control. A key risk to managing the sale of all these isotopes is the lack of control over, and knowledge of, their supply. Under the federal standards for internal control, federal managers are to assess the risks faced entity-wide, and at the activity level, from external and internal sources and decide what actions to take to mitigate such risks. Because Isotope Program officials do not believe that they have a stewardship role for helium-3, they did not take steps to mitigate the risk of selling helium-3 without information on the size of the inventory or its rate of replenishment. Similarly, without a stewardship role, the program did not take sufficient steps to accurately forecast the increased demand for helium-3. Such forecasting is important in order to align demand with current inventory levels. In this regard, the subcommittee report of the Nuclear Science Advisory Committee stated that for the Isotope Program to be efficient and effective, it needs accurate forecasts for the demand for isotopes. A lack of communication and failure to identify risks and forecast demand ultimately delayed the Isotope Program's awareness of, and the federal government's response to, the helium-3 shortage. In conclusion, we believe that all isotopes without clear stewardship responsibilities may face the same risks that led to the helium-3 shortage.

## RECOMMENDATIONS FOR EXECUTIVE ACTION

We are making four recommendations to the Secretary of Energy designed to avoid future shortages associated with managing all isotopes that the Isotope Program sells but whose supply it does not control, including helium-3. First, we recommend that the Secretary of Energy clarify whether the stewardship for all these isotopes belongs with the Isotope Program or elsewhere within the Department of Energy. Once the stewardship for these isotopes has been assigned, we further recommend that the Secretary of Energy direct the head of the responsible office(s) to take the following three actions:

- develop and implement a communication process that provides complete information to the assigned entity on the production and inventory of isotopes that are produced outside the Isotope Program;
- develop strategic plans that, among other things, systematically assess and document risks to managing the isotopes and supporting activities, such as not having control over the supply of these isotopes, and implement actions needed to mitigate them; and
- develop and implement a method for forecasting the demand of isotopes that is more accurate than the one that is currently used. In this regard, the actions taken should be consistent with the forecasting recommendation from the subcommittee report of the Nuclear Science Advisory Committee.

## AGENCY COMMENTS AND OUR EVALUATION

We provided a draft of this report to the Secretaries of Energy and Homeland Security for their review and comment. DHS had no comments on the findings of the draft report or our recommendations. In a written response for DOE, the National Nuclear Security Administration's Associate Administrator for Management and Budget stated that he understands our recommendations and can implement them but took exception to our characterization of the Isotope Program's mission. In the report, we state that the Isotope Program's mission is to produce and sell isotopes and related isotope services, maintain the infrastructure required to do so, and conduct research and development on new and improved isotope production and processing techniques, which was its mission from 2003 through 2008—the time during which the Isotope Program was selling helium-3 by auction and the helium-3 shortage occurred. In its response, NNSA explained that the Isotope Program has been working to clarify its responsibilities for isotopes since 2009. In this regard, the DOE fiscal year 2012 Congressional Budget request describes the Isotope Program's mission as that of producing and distributing isotopes that are not commercially available and distributing other materials as a service to DOE. However, according to NNSA's comments, the Isotope Program does not have the mission to be the steward of stockpiles of other materials and their byproducts, including helium-3. Neither DOE's fiscal year 2012 Congressional Budget request, nor NNSA's comments explain what entity does have stewardship responsibility for helium-3 and several other isotopes that are sold by the Isotope Program, but produced elsewhere. Regardless of how the Isotope Program defines its mission today, at the most crucial time when helium-3 should have been carefully managed in order to avoid the sudden awareness of the shortage, no one entity believed it had the responsibility to do so. As a result, the shortage of an isotope that is critical to national security, research, industrial, and medical applications went undetected until the supply reached a critical level. Our recommendations are intended to assist in avoiding such a problem with helium-3 and other isotopes in the future.

DOE also provided technical comments that we incorporated as appropriate throughout the report. DOE's comments on our draft report are included in appendix I.

As agreed with your offices, unless you publicly announce the contents of this report earlier, we plan no further distribution until 30 days from the report date. At that time, we will send copies to the appropriate congressional committees, Secretaries of Energy and Homeland Security, Administrator of NNSA, National Security Staff, and other interested parties. The report will also be available at no charge on the GAO Web site at http://www.gao.gov.

If you or your staffs have any questions about this report, please contact Gene Aloise at (202) 512-3841 or aloisee@gao.gov or Timothy M. Persons at (202) 512-6412 or personst@gao.gov. Contact points for our Offices of Congressional Relations and Public Affairs may be found on the last page of this report. GAO staff who made key contributions to this report are listed in appendix II.

Gene Aloise
Director
Natural Resources and Environment

Timothy M. Persons, Ph.D.
Chief Scientist
Director, Center for Science, Technology, and Engineering

# APPENDIX I. COMMENTS FROM THE DEPARTMENT OF ENERGY

Department of Energy
National Nuclear Security Administration
Washington, DC 20585

May 4, 2011

Mr. Gene Aloise
Director
Natural Resources and Environment
Government Accountability Office
Washington, DC 20458

Dear Mr. Aloise:

The Department of Energy (Department) and National Nuclear Security Administration (NNSA) appreciates the opportunity to review the Government Accountability Office's (GAO) report, *Managing Critical Isotopes: Weaknesses in DOE's Management of Helium-3 Delayed the Federal Response to a Critical Supply Shortage*, GAO-11-472. In response to a request made by the Subcommittee on Investigations and Oversight, Committee on Science and Technology, House of Representatives, GAO was asked to review the Department of Energy's (DOE) management of He-3 (He-3) to: (1) determine the extent to which the federal government's response to the He-3 shortage was affected by DOE's management of He-3; (2) determine the federal government's priorities for allocating the limited supply of He-3; and (3) describe the steps that the federal government is taking to increase the He-3 supply and develop alternatives to He-3. NNSA is responding on behalf of the Department.

The Department recognizes the need for improvements in managing He-3 and began to address this issue in 2008 via the Integrated Product Team (Team). As described in this report, He-3 allocation priorities have been established that have brought the supply and demand into a better balance. The result has been to defer the total exhaustion of US-origin He-3 from 2011 until the 2017-2018 timeframe, allowing the Team to develop alternative technologies to reduce the need for He-3, as well as secure additional supply.

We understand the GAO recommendations and can implement them. We are concerned, however, that the report does not accurately reflect the mission of the Isotope Program (IP), which in important aspects is determined by Congressional direction to both the Office of Science (SC) and the NNSA, and is not open to interpretation. To clarify the roles of both SC and NNSA regarding isotope production, in 2009, Deputy Secretary of Energy Dan Poneman directed SC and NNSA to develop a joint statement, describing the Department's approach to isotope research, development, and production to minimize risk or misuse. As a result, since the FY 2011 Congressional Budget request the Department has included a joint SC/NNSA statement on responsibility for isotopes in the SC budget Overview. A copy of the statement in the FY 2012 Budget Request is enclosed. While the IP has the mission to produce and distribute isotopes that are not commercially available and can distribute legacy materials as a service to the Department, it does not have the mission to be the steward of stockpiles of legacy materials and their byproducts. We believe the report, as written, may mislead the reader with an impression of the IP's mission and role as related to He-3.

Below are a couple of points that we feel need some clarification:

- Throughout the report, the GAO says that NNSA "produces" He-3. NNSA does not produce He-3; rather He-3 is extracted. While this may seem an issue of semantics, we feel it is important to note that we never manufactured He-3 for the purpose of commercial sales. Rather, NNSA extracted He-3 as part of our mission to maintain the U.S. nuclear weapons stockpile. Additionally, on Page 5 of the report states that He-3 is rare because it is only produced through the radioactive decay of tritium. While He-3 is currently only extracted from the radioactive decay of tritium it exists in very low concentration in natural sources, for example natural gas.
- With respect to using Lithium 6 (Li-6) as an alternative technology, we are fully aware of the need to avoid incurring the same issues with the U.S. Li-6 supply. We have taken steps to evaluate the potential demand using the most conservative estimates for Li-6. Should the decision be made to use Li-6, we will take all necessary steps to ensure there is an adequate supply to meet demand.

Further, as a result of the Team's discussions, and a Workshop held in August 2008 on the Isotope Program's initiative, the supply of other isotopes which are sold by the Isotope Program, but produced elsewhere, such as Li-6 mentioned in the report, is being addressed. In the case of Li-6, for example, Portal Monitor users within the USG were asked to estimate their long term needs, assuming that all their monitors would use Li-6 rather than He-3. As a result of that discussion, a Li-6 stockpile has been established to ensure supply of Li-6 for Portal Monitor use. We would be happy to provide GAO with documents showing our work in this area.

Also enclosed are specific comments to help clarify and improve the report in areas that may be confusing or misleading.

If you have any questions concerning this response, please contact JoAnne Parker, Director, Office of Internal Controls, at 202-586-1913.

Sincerely,

Gerald L. Talbot, Jr.
Associate Administrator
for Management and Budget

Enclosures

# Appendix II. GAO Contacts and Staff Acknowledgments

## Staff Acknowledgment

In addition to the contact named above, Ned H. Woodward, Assistant Director; Eric Bachhuber; R. Scott Fletcher; and Wyatt R. Hundrup made key contributions to this report. Kendall Childers, Nancy Crothers, Cindy Gilbert, Jonathan Kucskar, and Mehrzad Nadji also made important contributions.

## End Notes

[1] Tritium radioactively decays into helium-3 at an annual rate of 5.5 percent. For further information about tritium production, please see GAO, *Nuclear Weapons: National Nuclear Security Administration Needs to Ensure Continued Availability of Tritium for the Weapons Stockpile*, GAO-11-100 (Washington, D.C.: October 7, 2010).

[2] Congress created NNSA as a semiautonomous agency within DOE under title 32 of the National Defense Authorization Act for Fiscal Year 2000 (Pub. L. No. 106-65, § 3211, 113 Stat. 957 (1999)). NNSA is responsible for the management and security of the nation's nuclear weapons, nonproliferation, and naval reactors programs.

[3] The Isotope Program did not sell helium-3 from about 1995 through 2001 because helium-3 was being stockpiled for use in NNSA's Accelerator Production of Tritium project. During this time, Russia was the primary source of commercially available helium-3.

[4] Isotopes are chemical elements or varieties of a given chemical element with the same number of protons but different numbers of neutrons (e.g., helium-3 has one less neutron than helium-4, the helium isotope that is commonly used in party balloons).

[5] Linde acquired Spectra Gases, one of the principal buyers of helium-3, in 2006; Spectra Gases changed its name to Linde in 2009.

[6] GAO, *Standards for Internal Control in the Federal Government,* GAO/AIMD-00-21.3.1 ("Green Book") (Washington, D.C.: November 1999).

[7] The National Security Staff, established under the National Security Advisor, support all White House policymaking activities related to international and homeland security matters.

[8] Chronic obstructive pulmonary disease, commonly referred to as COPD, is a lung disease in which the lungs are partially blocked, making it difficult to breathe. In this type of research, a patient inhales helium-3 during the MRI so that doctors may get a clear view of the entire pulmonary structure.

[9] DOE, *Workshop on the Nation's Needs for Isotopes: Present and Future,* DOE/SC-0107 (Rockville, Maryland, December 2008).

[10] GAO/AIMD-00-21.3.1.

[11] Nuclear Science Advisory Committee, NSAC Isotopes Subcommittee, *Isotopes for the Nation's Future: A Long Range Plan. Final Report: Second of Two NSAC Charges on the Isotope Development and Production and Research for Applications Program* (August 27, 2009).

[12] Current good manufacturing practices (cGMP) are regulations set by the Food and Drug Administration for the manufacture of drugs, medical devices, and other products to ensure that companies' products meet specific requirements for identity, strength, quality, and purity.

[13] From 2003 through 2009, before helium-3 allocations were managed by the policy committee, the price for helium-3 ranged from about $40 per liter to about $85 per liter.

[14] Heavy-water reactors are nuclear power reactors that use water containing deuterium oxide as a coolant and natural (not enriched) uranium as its fuel source.

[15] We are currently conducting a technology assessment of alternatives for detecting neutrons. This technology assessment will examine the potential and maturity of various technologies that could lower demand for helium-3. We expect to issue a report on this assessment later in 2011.

In: Helium-3: Critical Uses and Supply Challenges ISBN: 978-1-61470-823-0
Editors: Gabriel A. Ross 

*Chapter 3*

# HEARING "CAUSES AND CONSEQUENCES OF THE HELIUM-3 SUPPLY CRISIS", BEFORE THE U.S. HOUSE OF REPRESENTATIVES, COMMITTEE ON SCIENCE AND TECHNOLOGY, SUBCOMMITTEE ON INVESTIGATIONS AND OVERSIGHT

***Gabriel A. Ross***

## PURPOSE

The Subcommittee on Investigations and Oversight meets on April 22, 2010, to examine the causes and consequences of the Helium-3 supply crisis. Helium-3 (He-3) is a rare, nonradioactive gas that has been produced in both the United States and Russia as a by-product of nuclear weapons development. Tritium, which helps boost the yield of nuclear weapons, decays into Helium-3 gas after approximately 12 1/2 years. The gas was produced as a consequence of tritium production by the defense programs of the Department of Energy (DOE). As a valuable commodity, it was packaged, managed and sold through DOE's Isotope Program in the Office of Nuclear Energy (though the Isotope program was moved to the Office of Science in a reorganization during FY2009).

# BACKGROUND

Helium-3 has wide-ranging applications as a neutron detector for nuclear safeguards, nonproliferation and homeland security purposes because it is able to detect neutron-emitting radioactive isotopes, such as plutonium, a key ingredient in certain types of nuclear weapons. Currently, almost 80 percent of its use is for safeguards and security purposes worldwide. It is also broadly used in cryogenics, including low-temperature physics; quantum computing; neutron scattering facilities; oil and gas exploration; lasers; gyroscopes; and medical lung imaging research.

During the Cold War, the U.S. had a steady supply of He-3 gas resulting from weapons production, but tritium production was halted in 1988. In the wake of the 9/11 terrorist attacks, however, the desire for radiation portal monitors and other nuclear detection equipment exploded. The Department of Homeland Security, for example, initiated a program to install more than 1,400 radiation portal monitors at ports and border crossings and also to supply smaller detectors to state and local governments. This enormous new demand came just as the available supply of Helium-3 was diminishing because of a reduction in nuclear weapons production. By early 2009, the total demand for helium was over 213,000 liters, and the supply was 45,000 liters.

The Department of Energy is the sole U.S. supplier of He-3 as part of its management of the nuclear weapons stockpile. They are also a key consumer of the gas because of their nuclear weapons detection program (the DOE Megaports and Second Line of Defense programs distribute PVT radiation portal monitors and other smaller detectors to nations around the world) and because of their support for spallation neutron sources. As the key supplier of He-3, as well as a consumer of the gas and a partner with agencies such as DHS and DOD in nuclear security, DOE was in a position to see the disconnect between an expanding demand and a declining supply. However, DOE failed to see the problem until the He-3 stockpile was nearly expended. This guaranteed that the He-3 shortage would become a crisis, rather than a smoothly managed transition to conserving and allocating supply to the highest use and obtaining alternative technologies.

It wasn't until late in 2008 that the Helium-3 supply shortage began to be identified as an issue by DOE when DNDO suppliers of He3 and other non-safeguards users could not obtain enough He-3 for their work. The last major allocation of He-3 had occurred in 2008 when DOE set aside 35,000 liters for the Spallation Neutron Source, an advanced neutron science research center at

DOE's Oak Ridge National Laboratory in Tennessee which the Department spent over $1 billion to construct.

By January of 2009, an inter-agency phone conference between DNDO and DOE occurred in which DOE established restrictions on the use of He-3. DNDO agreed to develop priorities for He-3 use and initiate a working group on the issue; DOE said it would start investigating alternatives. In the wake of that meeting, an interagency task force developed with participation by DNDO, DOE and the Department of Defense. That task force first met in March 2009. In the discussion that ensued, total annual government and non-governmental demand for FY2009 was projected as in excess of 213,000 liters. The total available stockpile was, at that time, just 45,000 liters. Out years show similar levels of demand while annual production was projected at 8,000 liters. As an appreciation of the scope of the problem developed among the key participants, other agencies were invited to participate. Work quickly began on allocation of He- 3 for FY 09 and 10, research on alternatives and investigation of possible sources of additional He-3, such as obtaining tritium from Candu reactors in Canada, Argentina and other countries to harvest He-3 and recycling and re-use of existing He-3. The entire process was "elevated" to the National Security Council when the DOD staffer heading up their He-3 effort was detailed to the NSC.

This process continued under the new Interagency Policy Committee (IPC), chaired by staff at the NSC. The Subcommittee has been told that allocation decisions for 2010 have been completed; the gas is now being processed and will soon be provided to those who have been approved to receive it.

## Impact of the Shortage

The domestic and global impact has been profound. The per-liter He-3 have skyrocketed from $200 to in excess of $2,000 per liter. (The Subcommittee has been told of one sale of Russian He-3 to a German firm at a price of $5,700 a liter.) The U.S. has essentially halted all exports of Helium-3 gas, and recently told the International Atomic Energy Agency (IAEA) that they will no longer be able to rely solely on the U.S. to provide them with He-3 gas for use in nonproliferation enforcement and verification actions. The Canadian government had to receive special permission from the U.S. prior to the Vancouver Olympics to permit the export of a He-3 mobile neutron detector for use at the Olympic Games.

For neutron scattering facilities that require tremendous amounts of Helium-3 gas, the situation is very grim. At least 15 of these multi-billion dollar research facilities are being or have been built in at least eight countries, including the U.S., United Kingdom, France, Germany, Switzerland, Japan, South Korea and China. By 2015, these facilities will require over 100,000 liters of He-3 gas, according to estimates provided to the Subcommittee. Most of those needs are unlikely to be met. There have been several international meetings of scientists discussing possible alternatives to He-3 for spallation neutron detection, but the research is in the very early stages.

Within the U.S. government, no program appears to have been more significantly affected than the Domestic Nuclear Detection Office's (DNDO's) Advanced Spectroscopic Portal (ASP) radiation monitor program, which relies on He-3 as its neutron detection source. The scale and scope of the Helium-3 crisis, however, and its impact on the ASP program in particular was not clearly known outside the government until the Investigations & Oversight Subcommittee held its second hearing on the ASP program on November 17, 2009. During that hearing, Dr. William Hagan, acting director of DNDO, testified that the Interagency Policy Committee had decided in September 2009 that He-3 would not be used radiation portal monitors. This was the first time the Subcommittee and the public were informed of the extent of the Helium-3 crisis. Surprisingly, even Raytheon, DNDO's prime contractor on the ASP program, did not become aware that a decision had been made to halt the supply of Helium-3 gas for their radiation portal monitors until they heard Dr. Hagan's testimony.

## SUMMARY

The shortage of He-3 was an inevitable consequence of a declining source from the U.S. nuclear weapons enterprise and a growing demand. However, the crisis and its jarring impacts were avoidable. With foresight on the part of DOE, the kinds of prioritization efforts now happening through the IPC could have started years ago. Research into alternatives to He-3 could have been well along to success, with some areas (such as portal monitor systems) lending themselves to alternatives more readily than others (cryogenics). In short, the stockpile could have been managed in a way that allowed for non-disruptive impacts to industry, researchers and the national security community. Instead, everyone is surprised and scrambling to identify alternatives, suspending their research and their production lines while hoping that a breakthrough in sources of He-3 or alternatives to He-3 happens very, very rapidly. The failure to manage the stockpile with an eye to demand, supply and future needs has had real consequences for many, many fields. Once the shortage became clear to all the key agencies, an interagency process that has laid out a rational guide to allocation and policies has emerged very quickly and appears to be well managed.

## WITNESSES

### Panel I

**Dr. William Hagan**, *Acting Director, Domestic Nuclear Detection Office (DNDO), Department of Homeland Security (DHS)*

**Dr. William Brinkman**, *Director of the Office of Science, Department of Energy (DOE)* (Dr. Brinkman will be accompanied by **Dr. Steven Aoki,** *Deputy Undersecretary of Energy for Counterterrorism and a Member of the White House He-3 Interagency Policy Committee (IPC) Steering Committee.*)

### Panel II

**Mr. Tom Anderson**, *Product Manager, Reuter-Stokes Radiation Measurement Solutions, GE Energy*

**Mr. Richard L. Arsenault**, *Director of Health, Safety, Security and Environment, ThruBit LLC*

**Dr. William Halperin**, *John Evans Professor of Physics, Department of Physics, Northwestern University*

**Dr. Jason C. Woods**, *Assistant Professor, Radiology, Mallinckrodt Institute of Radiology, Biomedical MR Laboratory, Washington University in St. Louis and Program Director, Hyperpolarized Media MR Study Group, International Society for Magnetic Resonance in Medicine (ISMRM)*

In: Helium-3: Critical Uses and Supply Challenges ISBN: 978-1-61470-823-0
Editors: Gabriel A. Ross 

*Chapter 4*

# OPENING STATEMENT BY CHAIRMAN BRAD MILLER, BEFORE THE U.S. HOUSE OF REPRESENTATIVES, COMMITTEE ON SCIENCE AND TECHNOLOGY, SUBCOMMITTEE ON INVESTIGATIONS AND OVERSIGHT, HEARING ON "CAUGHT BY SURPRISE: CAUSES AND CONSEQUENCES OF THE HELIUM-3 SUPPLY CRISIS"

***Brad Miller***

Five months ago, the Subcommittee held a hearing titled: The Science of Security: Lessons Learned in Developing, Testing and Operating Advanced Radiation Monitors. That hearing examined technical problems in the development of the Domestic Nuclear Detection Office's (DNDO's) new generation of radiation portal monitors called Advanced Spectroscopic Portals or ASPs. Among the issues the Subcommittee had expressed an interest in was the impact a reported shortage of Helium-3 was having on the ASP program.

At that hearing, Dr. Bill Hagan, the Acting Director of DNDO, (who joins us again today) testified that the shortage of Helium-3 was so severe that two months earlier a White House Interagency Policy Committee (IPC) had barred DNDO from using Helium-3 in radiation portal monitors. Since the

Department had not informed the Subcommittee of this situation, and the written testimony submitted to the Subcommittee also failed to make reference to the decision, we were surprised by the testimony. We were not the only ones to be surprised, among others taken by surprise was DNDO's the main ASP contractor, Raytheon.

What we have learned since is that both the Department of Energy and the Department of Homeland Security should have known several years ago that it would be a disaster to base radiation-detecting equipment on helium-3 technology. Helium-3 is a byproduct of tritium, and tritium's only purpose is to enhance the capability of nuclear weapons. Until recently, no tritium had been produced in this country since 1988, and the reduction in the nation's stockpile of nuclear weapons guaranteed a reduction in the stockpile of tritium – and helium-3.

After 9/11 – at the same time the supply was significantly decreasing – the demand for helium-3 grew exponentially for use in radiation detection devices It was also expanding for spallation neutron facilities worldwide, cyrogenic and medical research, and oil and gas exploration. The Department of Energy, which not only produces and sells helium-3, but is one of its largest consumers through the Megaports and Second Line of Defense programs and the Spallation Neutron Source at Oak Ridge, never – not once – warned anyone that there was no long-term supply for all of these uses, and they should begin looking for alternatives. In fact, in 2006, when there was only 150,000 liters left in the stockpile and many other users lined up, DOE told the Department of Homeland Security that there was enough for the 120,000 liters then estimated for the first phase of the ASP program. The result was that in mid-2008 when commercial vendors began to warn of a He-3 shortage, DHS didn't appear to have taken them seriously. It took several more months before there was government-wide acknowledgement of the severity of the problem.

The impacts of the helium-3 shortage are real and painful and extend well beyond Megaports, the Second Line of Defense and the ASP programs. Because of its unique physical properties, helium-3 plays a crucial role in oil and gas exploration, cryogenics (including low-temperature physics), quantum computing, neutron scattering facilities and medical lung imaging research. Important science is on hold in a wide range of fields and commercial opportunities for American firms that sell products using helium-3 have been lost. Over the past year the cost of obtaining Helium-3 has risen from around $200 per liter to more than $2,000 per liter.

The ongoing crisis has drastically delayed the ability of researchers and others to obtain helium-3 and prevented many firms and researchers from acquiring helium-3 at all, at any price. For many applications there are potential He-3 alternatives including boron-10 and lithium. For some work, particularly cryogenics-related applications, however, there are no known alternatives to using Helium-3 and these industries will need to continue to be supplied with He-3 if these industries and their scientific research programs are to continue.

Today, we will examine the causes and consequences of the Helium-3 supply crisis with a desire to learn lessons to guide future resource management. We also want to hear about the processes that are now in place to manage the limited supply of helium-3, to set priorities for access to that stockpile and the search for alternative sources and alternative gases. It is my understanding that allocations for 2010 have been determined, the gas is being processed and it will soon be distributed.

Looking back, it is clear that the shortage was inevitable. Helium-3 has been captured by the Department of Energy from the decay of tritium. With the end of the Cold War and the arms reduction agreements going back all the way to the Reagan Administration, the stockpile of tritium was not growing and so the production of Helium-3 would inevitably decline. Since 1991, DOE has allocated over 300,000 liters of helium-3, drawing the reserve down to a very low level by 2009. The annual production of Helium-3 from the U.S. tritium stockpile is now in the range of 8,000 liters per year and demand is orders of magnitude higher.

At the same time that production was declining, the demand for Helium-3 has been increasing since 9-11. Helium-3 has been a critical component in the portal radiation monitor programs at DHS and approximately 60,000 liters have been used in the current PVT systems alone. The ASP systems that Raytheon designed would have required, if a full acquisition had gone forward, approximately 200,000 liters of helium-3. The Department of Energy has its own radiation detection program in mega-ports with additional liters of helium-3 used in that program. Handheld and backpack radiation detection systems at DHS, DOE and also DOD are another ongoing source of expanded demand since 9-11.

In addition to this new security-related source of demand, the Spallation Neutron Source project, also a DOE program was moving towards conclusion, with its main detector requiring an additional 17,000 liters. With countries around the world all pushing to get into SNS-style research, the global demand in coming years for Helium-3 from these detectors alone is expected to exceed 100,000 liters.

Since the shortage was inevitable, does it matter that DOE failed to see that their stockpile was evaporating? Yes, it absolutely does matter. If DOE had noticed the disconnect between growing demand and declining supply, they could have managed the stockpile with clear prioritization for highest use, and led an aggressive and timely search for alternatives to helium-3. These actions would have helped us avoid this crisis. It is astonishing that DOE did not see this coming.

It also astonishes me that DNDO did not validate that sufficient resources of helium-3 were available for the ASP program. A cautious and reasonable analyst would have sought a complete accounting from DOE before wagering years of effort and hundreds of millions of dollars.

Good crisis management is an inspiring thing to see in the government and I have to say that the current efforts of DNDO, DOE, DOD and other agencies under the orchestration of the National Security Council staff appears to be very well organized. They have set out to do a thorough survey of demand and have attempted to identify all outlying sources of supply. They are identifying alternative gases and locating international opportunities to temporarily expand the supply of Helium-3. All of this is laudatory, and can serve as a nice model for future interagency management of crises, but even better is to avoid a situation requiring crisis management in the first place. I hope that DOE has learned a lesson with Helium-3 that will lead to wiser management of the unique isotopes they control and distribute.

The final lesson I hope the agencies and the White House learn is that when a Subcommittee asks for your documents, you have to produce them or explain why you cannot. The Subcommittee wrote to both the Department of Energy and the Department of Homeland Security on March 8 requesting materials by March 29. Neither agency responded in a timely fashion. Neither agency has produced all of their materials, nor offered anything approaching a comprehensible explanation of the situation. Allegedly, some small set of documents were originally produced by White House staff and distributed to the agencies, and I have been surprised at the difficulty of getting the White House and the agencies to simply do the reviews that the precedents of legislative-executive relations suggest should properly occur for these documents, which do not appear to rise to the level of an executive privilege claim. I am hopeful that we will break this impasse soon.

The implications of the situation are that the Subcommittee is not as prepared for this hearing as we should properly be. The agencies have gone through elaborate fictional inter-agency courtesies allowing for duplicative, time-consuming reviews. There is no legal basis for these reviews. This has not only wasted time but is discourteous to the Committee. As a result, it is my intention to leave the hearing record open and, in consultation with my Ranking Member, Dr. Broun, to include in the record relevant materials that are responsive to my original letter. I will not rule out a second hearing on this subject if the documentary record contradicts testimony we receive today nor would I rule out taking any other steps necessary to compel production of agency records. I hope it won't come to that, but I had enough of stonewalling and slow rolls by the last Administration to have much patience with it from this Administration.

In: Helium-3: Critical Uses and Supply Challenges ISBN: 978-1-61470-823-0
Editors: Gabriel A. Ross 

*Chapter 5*

# OPENING STATEMENT OF DR. WILLIAM K. HAGAN, ACTING DIRECTOR, DOMESTIC NUCLEAR DETECTION OFFICE, DEPARTMENT OF HOMELAND SECURITY, BEFORE THE U.S. HOUSE OF REPRESENTATIVES, COMMITTEE ON SCIENCE AND TECHNOLOGY, SUBCOMMITTEE ON INVESTIGATIONS AND OVERSIGHT, HEARING ON "CAUGHT BY SURPRISE: CAUSES AND CONSEQUENCES OF THE HELIUM-3 SUPPLY CRISIS"

*William K. Hagan*

## INTRODUCTION

Good morning Chairman Miller, Ranking Member Broun, and distinguished members of the Subcommittee. As Acting Director of the Domestic Nuclear Detection Office (DNDO) at the Department of Homeland Security (DHS), I would like to thank the Committee for the opportunity to discuss the helium-3 (He-3) supply.

As requested, my testimony today will address the following points:

- How we became aware of the shortage of He-3;
- How we responded to it;
- What was done at the beginning of the Advanced Spectroscopic Portal (ASP) program to ensure there was an adequate supply of He-3 to meet the program's needs;
- The impact of the shortage on DNDO's radiological and nuclear detection programs; and
- The status of the work we are doing to identify alternative technologies to replace He-3 as a neutron detector.

Since National Security Staff has recently briefed the Committee staff regarding the He- 3 shortage, I have limited my remarks today to DNDO actions related to He-3.

## HELIUM-3 SUPPLY

The United States' supply of He-3 has traditionally come from the decay of tritium, which the nation previously produced in large quantities as part of the U.S. nuclear weapons enterprise. The suspension of U.S. production of tritium in the late 1980s, however, resulted in a reduction in the amount of He-3 available for harvest. Currently, a significant portion of He-3 is used for neutron detection to aid in the prevention of nuclear terrorism. He-3 has become the overwhelmingly predominant technology used for this purpose; the Departments of Homeland Security, Defense (DoD), and Energy (DOE) each have nuclear detection programs that use He-3-based sensors. Additionally, He-3 is finding increasingly widespread use in areas beyond homeland security, including scientific research, medical, and industrial applications. Some of these applications may require relatively large volumes of He-3 for which there may be no known alternative. In the past, He-3 was a relatively low-cost commodity, and its use increased particularly with the advent of large radiation portal monitors both domestically and abroad. The limited supply of He-3, which is based on the nation's current stores of tritium, has been overwhelmed by this increase in demand. The current and future He-3 supply will fail to satisfy the demand of interagency partners and the commercial sector. Only approximately one tenth of the current demand for He-3 will be available from DOE/National Nuclear Security Administration

(NNSA) for the foreseeable future, and neutron detectors using He-3 are already becoming difficult to procure.

Since the inception of DHS in 2003, the majority of He-3 used was for the Radiation Portal Monitor (RPM) program. An RPM consists of a neutron detector, using He-3 gas in tubes, and a gamma detector, using large slabs of plastic scintillator. When DNDO was established in 2005, the RPM program was transferred from U.S. Customs and Border Protection (CBP). In FY 2006, when preparing to start a program for an advanced portal system, called the Advanced Spectroscopic Portal (ASP), DNDO met with DOE to discuss strategic resources that would be required for the ASP. DOE gave no indication that the supply of He-3 would be problematic, even with the amount of units we were envisioning.

Until recently, DHS acquired systems using He-3 by publishing an RFP and then reviewing responses to select a vendor or vendors. The bidders, in preparing their responses, would check the resources required to fulfill the order, including He-3. When this process was used at the beginning of the ASP program, none of the proposals indicated any issue with He-3 supply.

In the summer of 2008, DNDO first became aware of a potential problem with the He-3 supply through an email from a neutron detector tube manufacturer. Although DNDO investigated this issue, it was initially unclear whether the problem was a result of delays in the supply chain or an actual shortage of He-3. DOE, which traditionally has been responsible for managing and allocating the supply of He-3, issued a report verifying the existence and seriousness of the overall supply shortfall in the fall of 2008.

In February 2009, DNDO took the lead in forming the He-3 Interagency Integrated Product Team (IPT), with participation of DOE/NNSA and DoD, to assess the true impact of the shortage and to ensure that the most critical government and commercial programs would preferentially receive He-3. The IPT also began exploring opportunities to manage the existing He-3 stockpile; increase the supply of He-3; account for the entire demand for He-3; investigate alternative technologies to replace He-3 for neutron detection; adapt old technologies for retrofit into existing equipment; and examine policy issues that may impact the use, distribution, or production of He-3.

The IPT took steps to secure the He-3 necessary for high-priority programs, which included the RPM Program. DNDO also began negotiations in late January 2009 to secure He-3 for the ASP and other DNDO programs. This He-3 sale, which would have covered initial deployments of ASP, was finalized in June 2009. In July 2009, DNDO ceded control of this He-3 purchase to the National Security Staff Interagency Policy Committee to be allocated in accordance with interagency determinations in order to optimally satisfy the competing needs of He-3 users. As the He-3 is allocated to other agencies and departments, DNDO will be financially reimbursed. DNDO has continued to coordinate with interagency efforts to manage the He-3 shortage and actively participates in interagency working groups to address He-3 supply, demand, alternative technologies, and policy.

## IMPACT OF THE HELIUM-3 SHORTAGE

Because of the volume of He-3 required in the construction of RPMs and the desire to make sure that He-3 was being used for the highest interagency priorities, DNDO ceased to allocate any additional He-3 for RPMs in September 2009. Based on current funding and guidance for the RPM Program, the He-3 shortage has had no appreciable impact on the deployment of systems in FY 2010. The program has a sufficient inventory of RPM systems with He-3 tubes available to support deployments through FY 2011. Additionally, a number of solutions—including both the identification of new detector materials and management solutions to most effectively utilize existing supplies—are yielding results. If ASP units are certified for secondary scanning applications, DHS can reuse the He-3 from the existing RPMs that are being replaced and use it for the ASP units. Simultaneously, DNDO is leading interagency efforts to identify alternative neutron detectors that may eventually replace He-3 in these applications.

While other devices (for example, handheld radioisotope identification devices and backpack detectors used by the U.S. Coast Guard, CBP and the Transportation Security Administration) use smaller volumes of He-3, they are also impacted by this shortage. To mitigate the shortage and ensure supply to government customers, industry has been purchasing He-3 from other sources, such as private companies that have stored He-3, and recycling gas from obsolete equipment. This has offset some of the shortfall in the near-term, but a redesign of current equipment will be necessary over the next several years, once new neutron detection technologies have been identified. As such, DNDO plans to work with the device manufacturers to develop new technologies, integrate them into systems, and test them for suitability in the field. In the meantime, DNDO will also request modest allocations from the government stockpile to continue deployment of current human portable systems until alternatives are available.

## ALTERNATIVE NEUTRON DETECTION TECHNOLOGIES

As I mentioned earlier, the U.S. government is also exploring options to resolve this situation through the development of new types of neutron detectors. DNDO is at the forefront of these efforts and had been funding programs to address alternative neutron detection technologies as part of their mandate, prior to any knowledge of the He-3 shortage. We are also working with the interagency to engage the technical, commercial, and international communities to solicit ideas to address alternative materials for neutron detection. We are confident that the government, private industry, and international stakeholders are making progress on a prudent path forward. At present, we are working with the commercial sector to identify alternative detection products that have potential for near-term commercialization. Our DNDO Exploratory Research projects that address other detection materials with neutron capabilities have also been accelerated.

DNDO recently tested many known commercial off-the-shelf (COTS) and near-COTS alternatives for neutron detection and remains committed to working with the interagency to identify potential solutions. For RPMs that require large volumes of He-3, four technologies have been identified as being potentially viable candidates. Boron Trifluoride (BF3)-filled proportional counters were widely used for neutron detection before He-3-based detectors were available. DNDO conducted testing at a national laboratory to compare the performance of BF3 with the performance of He-3; while this testing validated the neutron detection capabilities of BF3 as a low cost replacement technology, we continue to seek additional alternatives because the hazardous material classification of BF3 makes it less attractive for end users.

Other promising technologies under development include Boron-lined proportional counters; Lithium-loaded glass fibers; coated non-scintillating plastic fibers; and a new scintillating crystal composed of Cesium-Lithium-Yttrium-Chloride, (Cs2LiYCl6) or CLYC, commonly pronounced "click", that has both neutron and gamma detection capabilities. Some of these new technologies may have neutron detection capabilities that meet or even exceed the abilities of current He-3-based detectors. Before any alternative is commercialized, we will check the availability of the key components to avoid another shortage issue.

Since He-3 was widely available and relatively inexpensive until only recently, alternatives are still somewhat early in their development, although these development efforts have been accelerated in the last year or so. DNDO will continue funding of exploratory research and early development, testing of new COTS and near-COTS alternatives, and acquisition of samples of promising technologies for more extensive testing and evaluation.

Chairman Miller, Ranking Member Broun, and Members of the Subcommittee, I thank you for your attention and will be happy to answer your questions.

In: Helium-3: Critical Uses and Supply Challenges ISBN: 978-1-61470-823-0
Editors: Gabriel A. Ross 

*Chapter 6*

# STATEMENT OF DR. WILLIAM F. BRINKMAN, DIRECTOR OF THE OFFICE OF SCIENCE FOR THE U.S. DEPARTMENT OF ENERGY, BEFORE THE U.S. HOUSE OF REPRESENTATIVES, COMMITTEE ON SCIENCE AND TECHNOLOGY SUBCOMMITTEE ON INVESTIGATIONS AND OVERSIGHT, HEARING ON "CAUGHT BY SURPRISE: CAUSES AND CONSEQUENCES OF THE HELIUM-3 SUPPLY CRISIS"

***William F. Brinkman***

Thank you Mr. Chairman, Ranking Member Broun, and Members of the Committee. I appreciate the opportunity to appear before you to provide testimony on the DOE's role and reaction to the national Helium-3 ($^3He$) shortage. Both the National Nuclear Security Administration (NNSA), and the DOE Isotope Development and Production for Research and Applications Program (Isotope Program) recently transferred to the Office of Science in the FY 2009 Appropriation, play a role in Helium-3 production and distribution. I have served as the Director of the Office of Science since June 2009, and I am pleased to share with you my perspectives on the role of the DOE Isotope Program in $^3$He production and distribution.

## OVERVIEW OF THE ROLE OF DOE IN HELIUM-3 PRODUCTION AND DISTRIBUTION

The DOE has supplied isotopes and isotope-related services to the Nation and to foreign countries for more than 50 years. Since its transfer to the Office of Science in 2009, the Isotope Program has continued to produce a suite of isotopes for research and applications that are in short supply, as well as technical services such as target development, chemical conversions, and other isotope associated activities. As part of this mission, the Isotope Program is responsible for the sale and distribution of $^3$He on behalf of DOE, but not for the production of $^3$He. $^3$He is a rare, non-radioactive and nonhazardous isotope of helium. Due to its low natural abundance, recovery from natural deposits has not been economically viable thus far. Instead, the sole production of $^3$He in the United States results from the refurbishment and dismantlement of nuclear weapons. The natural radioactive decay of tritium used in these weapons creates $^3$He, which is separated and stored during processing at the NNSA Savannah River Site (SRS) in South Carolina. To date, the only other commercial source of $^3$He has been from the decay of tritium that was produced within the former Soviet Union for its nuclear weapons program. Because the primary, current source of $^3$He is the decay of tritium, current supplies of this important gas are limited by the quantities of tritium on hand and being produced. Without development of alternative sources for $^3$He, use of this gas will be constrained seriously in the foreseeable future as accumulated stockpiles are drawn down.

The U.S. distribution of $^3$He for commercial consumption started in 1980. $^3$He production for commercial use, has never been a mission of the DOE. However, DOE made this byproduct of its operations available to scientific and industrial users at a price designed to recover extraction, purification, and administrative costs. Currently, the need for $^3$He in the United States is outpacing production.

The major application of $^3$He is for neutron detection, principally for national security purposes, nuclear safeguards measurements, oil and gas exploration, and in scientific experimentation. It is the preferred detector material for these applications because it is non-reactive/non-corrosive and it has the highest intrinsic efficiency for neutron detection. It is also important in low-temperature physics research and increasingly in medical diagnostics. A major use of $^3$He in U.S. research is for neutron detection in the Spallation Neutron Source (SNS), a one-of-a-kind, accelerator-based neutron source that provides intense pulsed neutron beams for scientific research, materials research, and industrial development. $^3$He is also used in dilution refrigeration in low-temperature physics experiments; there is no known alternative for this use.

The U.S. Government ceased reactor-based production of tritium for the nuclear weapons stockpile in 1988. Due to the downsizing of the world's nuclear stockpiles and the increase in the demand for $^3$He, we have reached a critical shortage in the global supply of $^3$He.

## REALIZATION OF $^3$HE SHORTAGE

From 1980 to 1995, $^3$He collected by the NNSA at the Savannah River Site (SRS) was purified at the Mound Laboratory along with other stable isotope gases for distribution by the Isotope Program. NNSA ceased operations at Mound, a laboratory used primarily for weapons research during the Cold War, in 1995. Between 1980 and 2003, the SRS had accumulated about 260,000 liters of unprocessed $^3$He. For security purposes, this total was closely held, and not known widely beyond DOE. Sales of this raw $^3$He by SRS began in 2003 as a remediation test project with the commercial firm, Spectra Gases (now named Linde LLC); Linde invested in excess of \$4,000,000 to establish purification capability of $^3$He. In August of 2003, NNSA and the DOE Office of Nuclear Energy, in which the Isotope Program resided at that time, entered into a Memorandum of Understanding for the sales of raw $^3$He derived from tritium processing. On October 2, 2003, the first invitation to bid on the sale of $^3$He was published in a FEDBIZOPS notice. There were three competitive auctions from 2003 until 2006. Some of the 2006 shipment occurred in 2007 and 2008. There were a total of 146,000 liters supplied primarily to two vendors. During this time period, the Isotope Program advised both vendors that the supply was limited to about 10,000 liters annually by NNSA. Between 2004 - 2008, an average of 25,000 liters of Russian $^3$He was entering the U.S. market annually. Since 2003, DOE has sold over 200,000 liters of $^3$He, drawing down a significant portion of the Department's inventory. In addition, allocations totaling 58,000 liters were provided to SNS directly from NNSA in 2001 and 2008 in support of the high priority neutron scattering basic research program.

In March 2006, Isotope Program was briefed by Systems Development and Acquisition, Domestic Nuclear Detection Office (DNDO) on the development and acquisition of the deployment of their domestic detection system. The goal was to award contracts by July 2006. There was discussion that additional $^3$He would be required by DNDO, but final quantities could not be provided at that time. Some quantities were discussed prior to the meeting, particularly taking into account the availability at the time of additional supply from Russia. In the fall of 2007, vendors expressed interest to the Office of Nuclear Energy Isotope Program about the timing of the next bid of $^3$He and the probability of increased needs, but actual quantities were not known. While it was becoming apparent that a gap between supply and demand was emerging the magnitude of the projected demand was still unknown, as was the future availability of $^3$He gas from Russia. A combination of $^3$He loading enhancements at SRS in 2007, which delayed $^3$He distribution capabilities, and a lack of detailed information on demand caused the planned 2007 bid to be delayed.

In 2008, concerned that the overall demand would surpass the available supply, even though the U.S. was not the sole source at the time, the Isotope Program delayed all further bid sales until additional information could be obtained. The Office of Nuclear Physics, in anticipation of the transfer of the Isotope Program from the Office of Nuclear Energy to the Office of Science, organized a workshop on the Nation's needs for isotopes for research and applications. This August 2008 workshop was attended by national laboratories, universities, industry, and federal agencies, including the Department of Homeland Security, and NNSA. At the workshop, the community discussed a demand for $^3$He approaching 70,000 liters annually'. The projected U.S. supply in the out years was estimated, at that time, to be about 8,000 liters annually. The results of the workshop were subsequently released in a report to the interagency community. During the same time period, Russia ceased offering $^3$He to the commercial market, informing U.S. vendors that it was reserving its supplies for domestic use.

## DOE Response to $^3$He Shortage

With the estimated magnitude of the shortage becoming clear in August 2008, the Isotope Program coordinated sales in 2008 among the Department of

Homeland Security (DHS), the NNSA Second Line of Defense (SLD) program, and industry, and did not distribute $^{3}$He through an open bid process. A briefing by the Isotope Program was held at DHS, with attendance by Department of Defense, DHS and NNSA, to discuss the projected $^{3}$He shortage. The DOE was instrumental in the development of the self-formed interagency group that was established in March 2009, with the objective of identifying the $^{3}$He demand and supply and R&D efforts on alternative technologies.

DOE quickly implemented a number of actions. NNSA and Office of Science agreed that no further $^{3}$He allocations would be made without interagency agreement. Together with DHS, they decided not to provide additional gas for portal monitor systems, which accounted for up to 80 percent of projected future demand. DOE accelerated plans for the development and deployment of alternative neutron detection technology to reduce demand, with the aim to begin implementation within the next few years. DOE started investigating the identification of new sources of $^{3}$He from other countries, including Canada, which could increase the domestic supply starting in two to three years. Together with DHS, DOE also started examining additional new $^{3}$He production from either natural gas distillation or new reactor-based irradiation. These options were seen as a long-term and expensive, but potentially necessary if demand continues to outpace supply in the future.

A targeted public outreach campaign was instituted to help ensure that the $^{3}$He user community was made aware of the current shortage. The DOE Isotope Program published the Workshop Report, which articulated the $^{3}$He shortage, and broadly disseminated the report to stakeholders and interested parties in December 2008. Both NNSA and the Office of Science made a formal inquiry in July 2009 to national laboratories and universities supported by their programs, explaining the shortage and asking for input on use, demand and alternatives. The public outreach campaign included letters to scientific associations involved in cryogenics, nuclear detection, medicine, and basic research, alerting them and their members of the shortage. Dedicated $^{3}$He sessions at technical association meetings such as the American Association for the Advancement of Science, National Academy of Sciences, American Nuclear Society, Institute of Nuclear Materials Management and Institute of Electrical and Electronics Engineers were arranged. The Isotope Program posted a fact sheet on the $^{3}$He shortage on both the Office of Nuclear Physics Website and the Isotope Business Office website in August 2009, notifying stakeholders of the shortage and informing them of the interagency efforts.

In July 2009, the White House National Security Staff (NSS) formed an Interagency Policy Committee (IPC), with broad federal representation, to investigate strategies to decrease overall demand for $^{3}$He, increase supply, and make recommendations to optimally allocate existing supplies. Both NNSA and the Office of Science are members of the IPC and the working groups that subsequently have been formed. The DOE, through its Isotope Program, presently is distributing the 2010 allocations of $^{3}$He to federal and non-federal entities, based on the recommendation of the IPC. The allocation process gives priority to scientific uses dependent on unique physical properties of $^{3}$He and to maintaining continuity of activities with significant sunk costs. It also provides some supply for non-government sponsored uses, principally oil and gas exploration. The Isotope Program is working closely with $^{3}$He industrial distributors to ensure that the available $^{3}$He is being distributed in accordance with the Interagency Working Group decisions.

Preliminary results obtained by the interagency group, projected FY 2010 U.S. demand to be 76,330 liters, far outpacing the total available supply of 47,600 liters or projected annual production of 8,000 liters. Based on guidance developed by the group, agencies have reduced their projected needs to 16,549 liters. A second review produced further reductions to 14,557 liters for FY 2010. At a December 10, 2009 meeting, the task force agreed to allocate a portion of this revised amount.

To achieve this reduction in demand, DHS and DOE have agreed to make no new allocations of $^{3}$He for use in portal monitors, which employ the largest quantities of this material in the allocation process. The NNSA Second Line of Defense program will continue carrying out its mission to deploy portal monitors, by using past allotments that provide sufficient $^{3}$He to support SLD activities through early FY 2011.

# IMPACT OF $^{3}$HE SHORTAGE

## International Safeguards

The current shortage has had the most severe impact on U.S. international safeguards efforts. Historically, due to the low cost of $^{3}$He, the U.S. has been the major supplier of $^{3}$He in support of International Atomic Energy Agency (IAEA) safeguards efforts. $^{3}$He is the neutron detector material in systems used for nuclear material accountancy measurements that help assure that nuclear materials have not been diverted. Except for the U.S. mixed oxide fuel (MOX) facility, which received its full request, all other U.S. international safeguards support is currently on hold as a result of the $^{3}He$ supply shortage. Concern about undermining the U.S. Government international safeguards efforts at the Japan MOX (JMOX) facility resulted in further investigation of international options for $^{3}He$ supply and verification of the operational timeline for JMOX. The IAEA is currently reaching out to Member States requesting they support JMOX by making $^{3}He$ available. The U.S. has offered to work with potential $^{3}He$ suppliers on extraction processes. NNSA's Office of Nonproliferation and International Security also has been working with Japan on an updated operational timeline. The original 2,800 liter request for FY 2010 has been scaled back to 1,000 liters and approved.

In the case of international safeguards, it is DOE's view that the shortage should not be viewed as just a U.S. problem, but rather one that will require international cooperation to solve. The U. S. has met with IAEA representatives, including Director General Amano, and has obtained full and active IAEA support for outreach to potential international suppliers. DOE also suggested that Russia provide $^{3}He$ from its reserves in support of these international safeguards efforts. The safeguards community both in the U. S. and internationally has reexamined its $^{3}He$ needs and the timing of those needs, with a view to phasing in installation of detectors that use non-$^{3}He$ technology, without negative impact to safeguards requirements.

## Second Line of Defense (SLD)

Portal monitors have been the largest use of $^{3}He$ in the past few years, accounting for about one-third of the total annual use. Given that most of the alternative development work is focused portal monitors, the IPC allocation process eliminated $^{3}He$ allocations for this use. Past FY 2011, this decision could potentially impact the SLD program.

SLD has a sufficient number of $^{3}He$-loaded detection tubes to complete its planned deployments through FY 2011. After that, SLD would be dependent on alternative technology for neutron detection. However, boron tri-fluoride (BF3), the neutron detection technology in use before $^{3}He$ became the preferred alternative, is toxic when exposed to air, leading to difficulties with handling, international shipping, and deployment of monitors in foreign locations. Several new neutron detection technologies are currently being tested by DHS and DOE. However, these need to be brought to full deployment readiness, married with portal technology, and formally tested by

SLD for detection capability and robustness, in accordance with the SLD mission and standards. It is estimated that two to three more years of development will be required before detection systems based on these technologies will be available for deployment.

## Other Users

$^3$He is used in support of lung imaging research. Constraining allocations or increased gas costs may have an impact on future pulmonary research efforts, particularly long term studies that use and provide historical data. For FY 2010, the medical community received 1,800 liters of gas which supports current activities. The medical research community is working with industry to recapture, recover and recycle $^3$He used for pulmonary research.

$^3$He is used as the refrigerant for ultra-low-temperature coolers for physics research, such as nanoscience and the emerging field of quantum computing. $^3$He is unique in that there are no materials other than helium that remain liquid at temperatures closely approaching absolute zero, and $^3$He's nuclear properties provide a handle to do cooling that $^4$He doesn't provide, allowing for cooling down to the milli-Kelvin level. In FY 2010, the full U.S. cryogenics request for 1,000 liters was approved. The true impacts to both R&D and operational programs will be better quantified in the upcoming months, as users with small volume requirements place orders for their projects.

$^3$He is a component of ring laser gyros, used in guidance and navigation equipment utilized by the DoD for strategic and tactical programs. These systems are utilized in guidance for smart munitions and missiles and in military aircraft and surface vehicle and navigation systems. They are also used in space guidance and navigation systems. $^3$He is required until current testing and qualification tests to assess an alternative gas are completed.

$^3$He plays an important role in basic research. Neutron scattering provides unique information about the structure and dynamics at the atomic and molecular level for a wide variety of different materials. Neutron scattering instruments have the requirements of high efficiency, very good signal-to-background ratio, and high stability of signal and background. Many neutron instruments depend on the use of $^3$He detectors because of their insensitivity to gamma rays, which permits measurements spanning very large dynamic ranges. They have high efficiency (>50%) for thermal neutrons, and their high stability permits precise measurements over long periods of time or with different sample conditions. No other detector technology currently comes close to matching these capabilities. A number of the neutron scattering instruments at the Office of Science High Flux Isotope Reactor (HFIR) and the SNS at ORNL already use $^3$He-based detectors. The shortage has not yet impacted the U.S. neutron scattering research community. It is projected that their $^3$He allocation will support experiments through FY 2014.

In addition, the international neutron scattering community is developing and installing new facilities that are projected to require approximately 120,000 liters of new $^3$He over the course of this decade. The U.S. neutron scattering community has been actively engaged with their international counterparts in investigating ways to reduce the total demand, make better use of available supply, and develop alternative technologies. The U.S. has insisted that international partners take responsibility for securing new sources of $^3$He, that the U.S. can no longer be the major supplier satisfying these needs.

## Alternative Sources of $^3$He

The DOE is pursuing multiple approaches to identify alternative sources of $^3$He.

### Reuse and Recycle

In the medium term (1-3 years), the focus is on investigating ways to increase and/or improve use of $^3$He supplies. DOE programs, such as the Emergency Response Program which uses backpack-sized $^3$He-based detection equipment for their nuclear search mission, and the international safeguards program have instituted recycle and recovery efforts. These efforts, have led to reductions in their overall demands for new $^3$He by about 10 percent. Other programs, such as SLD, have been able to reduce the total amount of $^3$He required in each system and still meet required specifications. The Office of Science also has been developing recycling approaches for its uses of $^3$He.

To help identify stray inventories of $^3$He, DOE/NNSA and Office of Science have issued a call to the laboratories and plants, directing that they inventory unused/excess bulk $^3$He quantities and equipment containing $^3$He. This could be used in the preparation of a DOE/NNSA recycling program that could be expanded to other government agencies. The DOE laboratories are analyzing the extraction process used to remove $^{3}He$ from tritium to determine if it can be further optimized. Savannah River National Laboratory is developing a process to extract $^{3}He$ from retired tritium equipment that otherwise would have been discarded. The process may provide as much as an additional 10,000 liters of $^{3}He$.

## New Supply

Tritium is produced by neutron capture in heavy-water-moderated reactors, such as those used in Canada, Argentina and other countries. Because tritium is radioactive, utilities using these types of reactors often need to separate and store tritium in sealed containers, where it decays to produce $^{3}He$. Typically these containers have been designed to support permanent storage, not future extraction. DOE/NNSA is discussing with these countries how much, if any, $^{3}He$ they have in storage and how best to secure and make available. Investigations into possible ways to secure that material include transporting the storage containers to the U.S. for extraction in the U.S. or licensing the U.S. extraction process at the foreign facility. These are on-going negotiations; additional details can be provided once agreements have been reached with potential partners. Based on preliminary estimates, DOE/NNSA believes it would be possible to extract approximately 100,000 liters of $^{3}He$ over a 7-year period. The results of technical feasibility and cost studies are expected to be available by early FY 2011 as a basis for decisions by DOE and other interested agencies.

Over the longer term, it may be possible to produce $^{3}He$ rather than derive it as a byproduct of other activities. DOE/NNSA is currently examining the feasibility of two possible pathways. However, both of these options would require capital investment by DOE or another agency, and would likely involve a substantial increase in the cost of $^{3}He$ to the end user.

First, it may be possible to extract $^{3}He$ from natural gas. A 1990 Department of Interior (DOI) Study entitled, "Method and Apparatus for Direct Determination of $^{3}He$ in Natural Gas and Helium" found wide variations in the amount of $^{3}He$ at various drilling sites, ranging from less than 1 part per billion to over 200 parts per billion.

Secondly, the NNA Office of Defense Programs is evaluating the cost and feasibility of conducting reactor-based irradiations to produce tritium for the primary purpose of subsequent $^{3}He$ harvesting. This approach would utilize the facilities currently employed to generate tritium for the nuclear weapons stockpile. Although the necessary infrastructure currently is in place, additional costs would be incurred for target fabrication and subsequent processing. Because of the 12.3-year half life of tritium, there would be a delay of a number of years before any new $^3$He would become available.

### Non $^3$He Based Detectors

In FY 2009, NNSA initiated a program to address the shortage of $^3$He that focuses on non-$^3$He replacement technologies for neutron detectors in portal monitors deployed by the SLD Program. The NNSA Office of Nonproliferation and Verification Research and Development has, for many years, been developing alternative neutron detection technologies, but these efforts were not focused on portal monitoring applications that require large-area detectors. Since FY 2009, this application has become the principal focus of this neutron detection R&D program. Several promising technologies are being investigated that could supplement the use of the older BF3 technology as substitutes for $^3$He neutron detectors.

## Current Actions and Allocation Process for Helium-3

The NSS IPC met in September 2009 and concurred on a strategy that decreases overall demand for $^3$He, including conservation and alternative technologies, increases supply through exploring foreign supplies/inventories and recycling, and optimally allocates existing supplies. Furthermore, the IPC agreed to defer all further allocation of $^3$He for portal monitors, beginning in FY 2010, and would not support allocating $^3$He for new initiatives that would result in an expanding $^3$He infrastructure. The IPC stipulated that $^3$He requests should be ranked according to the following priorities:

1. programs requiring the unique physical properties of $^3$He have first priority.
2. programs that secure the threat furthest away from US territory and interests have second priority.
3. programs for which substantial costs have been incurred will have third priority.

Adoption of this approach for managing the U.S. $^{3}$He inventory produces allocations for Fiscal Years 2010 through 2017 that can be met by projected reserves. This is in contrast to the original allocation approach, which would have resulted in large and increasing shortages over the same period of time.

For FY 2010, allocations were as follows:

| | | |
|---|---|---|
| a. | DOE (Safeguards) | 800 liters (+ 1000 liters)* |
| b. | DOE (Detection) | 1,520 liters |
| c. | DOE (Emergency Response) | 1,750 liters |
| d. | DOE (NIF/NNSA) | 80 liters |
| e. | DOE-Science | 341 liters |
| f. | NIST | 832 liters |
| g. | Oil and Gas | 1,000 liters |
| h. | NIH (Med Imaging) | 1,800 liters |
| i. | Cryogenics | 1,800 liters |
| j. | NASA | 80 liters |
| k. | Environ Management | 0 liters |
| l. | IC | 0 liters |
| m. | DoD | 882 liters (+648 liters)** |
| n. | DHS | 772 liters |
| o. | DOS | 100 liters |

*DOE requested and was approved for an additional 1000 liters for the JMOX facility in FY10.

**DoD requested and was approved for an additional 648 liters in FY10. 325 liters will be used for the guidance and navigation systems, and 323 liters will be used by the DoD laboratories for cryogenic dilution refrigeration.

## CONCLUDING REMARKS

The DOE is committed to working with other agencies, the community and the White House in reducing the demand of $^{3}$He, increasing the supply of $^{3}$He, and distributing $^{3}$He in accordance to the Nation's highest priorities.

Thank you, Mr. Chairman and Members of the Committee, for providing this opportunity to discuss the national $^{3}$He shortage and DOE's roles and reaction to the shortage. I'm happy to answer any questions you may have

In: Helium-3: Critical Uses and Supply Challenges ISBN: 978-1-61470-823-0
Editors: Gabriel A. Ross 

*Chapter 7*

# WRITTEN TESTIMONY OF THOMAS R. ANDERSON, PRODUCT LINE LEADER, GE ENERGY, REUTER STOKES RADIATION MEASUREMENT SOLUTIONS, BEFORE THE U.S. HOUSE OF REPRESENTATIVES, COMMITTEE ON SCIENCE AND TECHNOLOGY, SUBCOMMITTEE ON INVESTIGATIONS AND OVERSIGHT, HEARING ON "CAUGHT BY SURPRISE: CAUSES AND CONSEQUENCES OF THE HELIUM-3 SUPPLY CRISIS"

***Thomas R. Anderson***

Mr. Chairman and members of the Subcommittee, my name is Tom Anderson and I am the Product Line Leader for GE Energy's Reuter Stokes Radiation Measurement Solutions. I appreciate the opportunity to testify before this Committee today.

I have been asked to speak about the impact the Helium-3 shortage has had on our business and our customers, and to share with the Committee our ideas on how to manage this problem in the future.

GE Energy's Reuter Stokes has over 50 years of experience supplying radiation detectors. We design and manufacture detectors for Boiling Water Reactors (BWR), neutron scattering instruments, oil and gas exploration, homeland security and nuclear safeguards systems. Our BWR in-core detectors monitor reactor power levels and provide signals to initiate protective actions in the event of an abnormal condition. Our Helium-3 gas-filled neutron detectors are used to accurately account for nuclear materials during handling and processing. Over 35,000 GE Helium-3 detectors are installed in systems deployed around the world today to monitor for the illicit trafficking of smuggled nuclear materials. I look forward to providing you with GE's perspective on the consequences of the Helium-3 supply crisis.

According to information presented at the Helium-3 Workshop hosted by the American Association for the Advancement of Science on April 6, 2010, the Department of Energy's Helium-3 reserves have been depleted to approximately 50,000 liters, with future production rates expected to be less than 10,000 liters per year. With global demand now on the order of 70,000 liters per year, the total DOE reserve represents less than a one-year supply of Helium-3. As a consequence, GE is confronting the reality that Helium-3 for use in neutron detectors may soon no longer be available.

In my testimony, I will address two points. First, a drop-in replacement technology for Helium-3 does not exist today. Furthermore, as many as six different neutron detection technologies may be required to best address the performance requirements of the neutron detection applications GE has served historically with technology using Helium- 3. Significant research is required immediately, and Federal funding is essential to accelerate development of new neutron detection technologies, and thereby preserve the remaining Helium-3 supply for other uses. Second, an adequate supply of Helium-3 must be made available by DOE and the Interagency Project Team (IPT) to support critical applications such as nuclear safeguards, homeland security and oil exploration while alternate technologies are developed.

## Background

GE Energy's Reuter Stokes business is located in Twinsburg, Ohio. Beginning with our first gas-filled neutron detector in 1956, GE has become a global leader in designing and manufacturing gamma and neutron detection technologies for a wide variety of applications.

Many of the Boiling Water Reactors (BWR) in operation in the United States today rely on GE detectors to measure and monitor reactor power level. Several U.S. states, as well as South Korea and Taiwan, have installed networks of Environmental Radiation Monitors manufactured by GE to monitor low-level gamma radiation.

GE also manufactures a variety of products for use in the oil and gas drilling and logging industry. These include sophisticated instruments to navigate a drill string; gamma radiation detectors to determine the type of rock and formation density; resistivity tools to measure formation properties and Helium-3 neutron detectors to measure formation porosity. The data from this full suite of detectors is integrated to optimize oil exploration.

During its long history, GE has designed and manufactured an assortment of BF3, Boron- 10 lined, and Helium-3 gas-filled neutron detectors. Over 100,000 of our Helium-3 neutron detectors have been put in service during the past four decades. Our neutron detectors have been utilized in a wide variety of neutron scattering research, nuclear safeguards, oil and gas, and homeland security systems.

Recently in the media, there has been much excitement and speculation about the presence of water on the Moon and on Mars. Our Helium-3 detectors have been used for space exploration where the unique properties of Helium-3 support water exploration at temperatures approaching absolute zero.

GE purchases the majority of its Helium-3 gas from the Department of Energy. The Helium-3 is processed and then used to manufacture Helium-3 neutron detectors. Our company does not otherwise bottle or package Helium-3 for sale.

The following sections provide background on four of the larger applications that use Helium-3 neutron detectors.

## Neutron Scattering Research

Neutron scattering facilities conduct fundamental science, materials, electromagnetics, food and medical research by directing a beam of conditioned neutrons at a test specimen and accurately measuring the position and timing of the scattered neutrons. GE is the industry leader in engineering and manufacturing Helium-3 gas-filled, position- sensitive neutron detectors for neutron scattering research facilities located around the globe. The three largest facilities in the United States are the Spallation Neutron Source (SNS) located at Oak Ridge National Laboratory, the National Institute of Standards and Technology (NIST) Center for Neutron Research (NCNR) in Gaithersburg, MD and the Los Alamos Neutron Science Center (LANSCE) located at Los Alamos National Laboratory (LANL). International facilities include the Japan Proton Accelerator Research Complex (JPARC), Rutherford Appleton Laboratory (UK), and Institut Laue-Langevin (France) as well as facilities located in Germany, South Korea, the Netherlands, Australia, and China. The research conducted at neutron scattering facilities has led to a long list of landmark discoveries including a better understanding of neurological and genetic diseases such as Huntington's disease, potential improvements in solar energy conversion, and advances in superconducting materials, to name but a few.[1]

Neutron scattering facilities represent a significant government research investment. The majority of the construction budget is used to build the neutron source, the accelerators and the infrastructure needed to support the scattering instruments. The construction cost for the SNS facility was $1.4 Billion.[2] The design and construction of the individual scattering instruments, including the Helium-3 detectors, is typically among the last tasks to be completed. The instrument arrays vary in size from tens of detectors to over 1,000 Helium-3 detectors per instrument. Instrument construction at many scattering facilities located outside the United States is currently on hold due to the lack of Helium-3.

Neutron scattering instruments require detectors with extremely fast response, high neutron sensitivity and excellent gamma discrimination. The detectors must provide accurate position and timing information for the scattered neutrons.

## NUCLEAR SAFEGUARDS

The purpose of nuclear safeguards programs is to prevent diversion of nuclear materials for non-peaceful purposes. Nuclear safeguards systems are installed at facilities that process, handle, use and store plutonium, uranium, nuclear fuel, spent fuel or nuclear waste. Safeguards systems quantify and monitor nuclear material to enable facilities to precisely account for plutonium and uranium during all aspects of processing, storage and clean up. The International Atomic Energy Agency (IAEA) and the National Nuclear Security Administration (NNSA) via the National Laboratories sponsor a number of international safeguards programs such as the new reprocessing facility that is under construction at the Rokkasho Reprocessing Complex in Japan.

Nuclear safeguards systems are typically compact. The detectors must have high neutron sensitivity and excellent gamma discrimination to enable accurate neutron measurements. The extremely fast response of Helium-3 detectors makes certain measurements possible. Helium-3 detector performance can be further tailored to permit highly precise nuclear material assay. This is a key element in accurately accounting for nuclear materials.

## OIL AND GAS

Helium-3 neutron detectors are also widely used in oil and gas exploration. These detectors are used in conjunction with a neutron source to locate hydrogenous materials such as oil, natural gas, and water. Neutron measurements in conjunction with inputs from other drill string instruments are used to locate hydrocarbon reservoirs during drilling, and to further delineate the reservoirs during logging operations. The overwhelming majority of nuclear porosity tools used in the oil and gas industry today depend on the unique properties of Helium-3 neutron detectors.

Helium-3 neutron detectors have high neutron sensitivity, which enables them to be packaged to fit inside the tool string. The excellent gamma discrimination characteristic of Helium-3 means that background gamma radiation levels do not interfere with the accuracy of the neutron measurements. These detectors must also operate reliably and survive at temperatures up to 200°C under severe vibration and shock levels up to 1,000 times the force of gravity. It is likely that without Helium-3, exploration for new reserves, development drilling of existing fields, and logging of both new and existing wells will be severely curtailed until an alternative technology is developed.

## HOMELAND SECURITY

The demand for Helium-3 neutron detectors has increased significantly since 9/11. Helium-3 is used as a neutron detector technology throughout the full spectrum of homeland security instruments, ranging from small 3/8" diameter detectors installed in pager-sized systems to six-foot long detectors installed in large area Radiation Portal Monitors (RPM). GE's Helium-3 detectors are widely used in radiation pagers, handheld instruments, fission meters, backpacks, mobile systems and RPMs that are deployed to search for and detect the illicit trafficking of fissile radioactive materials. Homeland security systems, particularly the RPMs, require a significant amount of Helium-3.

GE's Helium-3 neutron detectors are installed in systems supporting Customs and Border Protection (DHS), the Second Line of Defense (SLD)/Megaports Program (DOE) and the Advanced Spectroscopic Portal (ASP) Program (DHS). We have also manufactured thousands of Helium-3 detectors for other DHS, DOE (NNSA), Department of Defense (DoD), Department of Justice (DOJ), and other local and state security programs.

## Helium-3 Supply Concerns

The Department of Energy has been selling isotopes for several years. In December 2003, the DOE auctioned 95,800 liters[3] of Helium-3. An additional 50,848 liters were auctioned between 2005 and 2006.[4] After the last auction sale of Helium-3 in July 2006, there were repeated delays in the periodic auction process. In May 2008, GE met with the DOE to request clarification on the next anticipated auction date. It was during this May 2008 meeting that GE first became aware of the potential shortage of Helium- 3. In July 2008, the Department of Homeland Security's Domestic Nuclear Detection Office (DNDO) and the NNSA were briefed on the possibility that future supplies of Helium-3 might be inadequate to fully support their programs.

DOE suspended the anticipated 2008 auction and in December 2008 made a direct allocation of approximately 23,000 liters of Helium-3 to GE and Spectra Gases, Inc. Seventy percent of the Helium-3 sold to GE was controlled by NNSA for the Second Line of Defense (SLD) Program. There has been no additional Helium-3 auctioned by the DOE, and since 2008, all DOE gas supplied to GE has been allocated to specific projects or programs.

The impact of the Helium-3 shortage was immediate. GE was no longer able to supply products to many programs and customers. The neutron scattering community has been hardest hit, with programs in Japan and Germany having the most immediate need. The construction of several scattering instruments outside the United States will be delayed until a source of Helium-3 can be identified or an alternate technology is made available.

Upon learning of the Helium-3 shortage, GE designed and built equipment to more efficiently reclaim Helium-3 from unused detectors. Helium-3 is a stable gas, and therefore can be removed from old detectors, reprocessed and used to build new detectors. Recycled Helium-3 has been used over the past year to build neutron detectors for some systems.

## ALTERNATIVE TECHNOLOGIES

A drop-in replacement for Helium-3 does not exist today. Federal research funding is essential to supplement private sector efforts to accelerate development of replacement technologies. I have discussed four applications that currently rely on Helium-3 neutron detectors. I have also briefly described the detector performance attributes required in each. Many of the applications share similar attributes, yet each has its own subtle differences. Up to six different neutron detection technologies may be required to replace Helium-3 detectors in these four applications.

Three different technologies may be needed to support homeland security systems alone. The systems deployed for homeland security today range in size from large area portal systems and lightweight backpack instruments, to low-power pager-sized equipment. Neutron scattering detectors are even more complex due to the speed, timing and position measurement accuracies needed to support their research. Alternate technologies for nuclear safeguards and the extremely harsh conditions encountered during oil exploration also present unique development challenges.

GE has been actively involved in developing alternate neutron detection technologies. GE's initial efforts have been focused on developing a replacement technology for portal monitors. RPMs have been the largest consumer of Helium-3 during the past seven years. GE recently completed development of a Boron-10 lined gas-filled neutron detection technology that meets the American National Standards Institute (ANSI), ANSI N42.35-2006 performance requirements for portals. This was an accelerated project, which from initial concept to first production is on track to be completed in 18 months. For this project, our Twinsburg team worked with scientists at the GE Global Research Center and leveraged production processes based on best practices from GE Consumer and Industrial businesses. GE is on schedule to begin production of Boron-10 lined neutron detection portal panels in July of this year.

The research and new product development programs for the four neutron detection applications described will be challenging. Each new technology must the reliable and consistently meet the performance requirements needed for accurate neutron measurements under all system operating conditions. The technology must be scalable to fit the instrument and have a reasonable service life. Finally, the technology must be practical to manufacture in sufficient quantities at a reasonable cost, with consistent quality and performance.

GE is well qualified to research and develop new neutron detection technologies. However, research and development programs of this scope are very expensive. DNDO has released Broad Agency Announcements (BAA) and a Request for Information (RFI) to seek information and provide funding for alternate neutron detection technologies for homeland security systems. I am not aware of similar programs at DOE. Nuclear safeguards, oil exploration, and neutron scattering facilities fall under different offices within DOE. Federal funding to support research in each of these areas is needed if replacement technologies are to be in place in time to avoid serious effects of the Helium-3 shortage.

## Alternate Sources of Helium-3

Helium-3 is generated from the radioactive decay of tritium. During the Cold War, both the United States and Russia produced tritium to support nuclear weapons stockpiles. Most of the Helium-3 available today was harvested from the tritium produced for the weapons program.

Tritium is also produced as a byproduct of generating power in CANada Deuterium Uranium (CANDU) reactors. Four such reactors are located at Ontario Power Generation's (OPG) Darlington Generating Station in Ontario, Canada. GE has investigated the possibility of separating the Helium-3 from the tritium that is currently being stored at the Darlington facility. GE has been informed that the U.S. Government has initiated discussions with the Canadian government. If such discussions lead to an agreement, this might provide some additional Helium-3 to support critical applications while alternate technologies are developed.

## CONCLUSION

We have come to rely on Helium-3 for cutting-edge research, medical lung imaging, cryogenic cooling, oil and gas exploration, and the radiation monitors that protect our borders. The Department of Energy's Helium-3 reserve is nearly depleted and there are no short-term solutions available to rectify the shortage. An Interagency Project Team has been established to manage the shortage and to make the difficult decisions to allocate the remaining limited supply of Helium-3.

DNDO has played a key role in addressing the shortage, however, there is much more to be done. It is critical that the federal government strengthen its support of research and development for alternate technologies. Specifically, DOE funding of research and development programs for oil and gas exploration, neutron scattering and nuclear safeguards is essential. Funding and collaboration with the National Laboratories could help accelerate technology development. Also, additional funding from DNDO would help accelerate development of technologies for homeland security. Finally, it is extremely important that the Interagency Project Team allocate adequate supplies of the remaining Helium-3 to support critical applications such as oil exploration and nuclear safeguards while alternate technologies are developed. Given the limited Helium-3 supply, the Federal government should consider moving forward on negotiations with the Canadian government so that Helium-3 can be produced from the tritium currently being stored at the CANDU Darlington facility. This is not a long-term solution, but it may help provide a supplemental supply of Helium-3 while alternative solutions are found.

Thank you for holding this hearing on this critical issue. I will be glad to answer any questions you may have.

## End Notes

[1] Additional information is available on the Oak Ridge National Laboratory website: http://neutrons

[2] *Id.*

[3] Invitation for Bids to Purchase He-3 gas, Amendment 2, posted November 20, 2003.

[4] US DOE Helium-3 (He-3) Sales Solicitations (2005, 2006).

In: Helium-3: Critical Uses and Supply Challenges ISBN: 978-1-61470-823-0
Editors: Gabriel A. Ross 

*Chapter 8*

# TESTIMONY OF RICHARD L. ARSENAULT, THRUBIT LLC, HOUSTON, TEXAS, BEFORE THE U.S. HOUSE OF REPRESENTATIVES, COMMITTEE ON SCIENCE AND TECHNOLOGY, SUBCOMMITTEE ON INVESTIGATIONS AND OVERSIGHT,HEARING ON "CAUGHT BY SURPRISE: CAUSES AND CONSEQUENCES OF THE HELIUM-3 SUPPLY CRISIS"

*Richard L. Arsenault*

## INTRODUCTION

Chairman Miller, Ranking member Broun, and members of the Committee, my name is Richard Arsenault and I am the Director of Health, Safety, Security and Environment along with being the Corporate Radiation Safety Officer for ThruBit LLC (ThruBit Logging Solutions) which is a Shell Technology Ventures Fund 1 BV Portfolio company formed in 2005. Today we offer complete logging solutions based on a unique patented "through the bit" deployment technique that provides significant advantages in many applications. We are a small company taking this new technology from proof of concept to commercial introduction with aspirations to grow into a much larger company. I have been involved in the Oil Well Logging industry since 1979 starting out as an Open Hole Wireline Engineer in West Texas and later got involved in the early stages of Logging While Drilling in 1982.

## Well Logging

Every well requires formation evaluation; well logging is a key part of this evaluation. The quality and accuracy of data is key to decide and ascertain if the well is a producer or dry hole. This evaluation supports and drives:

- Production Estimations,
- Well Economics,
- Reserve calculations
- Corporate and Government Energy Assets,
- Overall market fundamentals

It supports ability to commit to long term projects with less than certain payback. Provides support for filing Company's statement of reserves. Helps value royalty payments back to state and federal government and drives legislation.

The US is most affected:

- 1/2 of worlds activity
- 1/4 of world consumption
- < 5% of world reserves
- Greatest need for immediate continuity of supply

## Neutron Logging

Wells can be logged by Wireline Logging or Logging-While-Drilling (LWD). There are a number of formation measurements that are taken when a well is logged. Neutron logging is one of the primary measurements taken when a well is logged. The neutron measurement provides the hydrogen located in the pore space of the formation and the porosity is determined from neutron counting rates in the detectors within the logging tool. The neutron measurement is a primary gas indicator which helps delineate gas and oil producing zones along with providing the porosity of the formation.

Both Wireline and LWD tools will in most cases have a “Long Space” and “Short Space” Helium-3 Detector which are located at different distances from the radioactive sources mounted in the logging tool. The Helium-3 detectors are used with either an Americum-241 Beryllium or Californium-252 radioactive source.

The importance of Helium-3 supply to the oil and gas industry is critical and crosses into numerous sectors of the industry. Helium-3 gas is used in almost the entire neutron detectors incorporated into downhole tools in our industry. The neutron count rate measurement, from which the porosity measurement is derived, is used in all oil and gas reservoir evaluations. Even small errors in the neutron measurement can make the difference in whether a reservoir is commercially viable or not.

It is difficult for our industry to determine the number of neutron detectors used in our course of business, especially since the neutron detector is used in open and cased hole compensated neutrons, single detector neutrons and other devices in our industry. There are numerous large well logging companies in the U.S. that also operate internationally along with medium to small size companies throughout the U.S. Each of these companies incorporates the use of He-3 neutron detectors in their tools. With the downturn in our industry over the last two years, most existing companies have been able to utilize existing tool stocks for replacement detectors and spare parts, which have lessened the impact over these years, but will eventually deplete the stock within those companies. They will be forced to buy additional detectors as the industry expands, for both new tools and for replacements in older tools. The detectors do have a limited life expectancy on the average of about 5 years depending on the downhole conditions they are exposed. So they do need to be replaced periodically to keep the tools working correctly. Companies introducing new technologies for logging wells, such as ThruBit, are limited to what is already available in house to build tools and what they can find available by the detectors suppliers with long leads time and a substantially higher price.

## Pricing and Availability of He-3 Detectors

We have personally seen almost a 3 times price increase and a quoted lead time of almost 6 months for delivery in an order recently placed this year. I have also received reports from others in the industry of pricing increases reported on neutron detectors in the 3 to 10 times range due to the Helium-3 shortage. Pricing is not the only issue, but availability is also key. Lead times of 6-8 months have been reported. There have been reports of some detectors not being available due to the lack of Helium-3.

There is a big difference in application of detector technology to applications that are located on surface, exposed to ambient temperatures and pressures and are not moved or exposed to conditions involving shock and vibration. Detector technology used in down hole tools used for well logging are subjected to more stringent requirements just to survive the environment and meet the engineering requirements of the design.

Wireline Tools are operated at high temperature, have limited internal geometry to mount the detectors and experience medium shock and vibration. In the case of LWD tools they have all the same factors, but the shock and vibration is a lot higher. As result of the limited internal geometry small reliable detector packages are a must. In our particular case we have the smallest well logging tools in the industry with a 2-1/8" diameter tool. Any type of alternative technology would require the same or smaller foot print inside the tool. We could not go larger since we limited to our 2-1/8" diameter specification. We do not have the resources for an R&D effort to pursue another tool design with potential alternative detector technology.

## Impact

Being a small company bringing new technology to market is a challenge. We are in transition from a commercial introduction phase to comercialization with an aggressive plan to be a full blown viable and sustainable Formation Evaluation Service Company. The Helium-3 detectors are all we have to put in our Neutron Porosity tools. We do not have a substitute detector for use in these well logging tools. It would take substantial development time (years) to pursue a substitute. We have neither the financial resources or R&D staff to pursue this effort. An extreme shortage or unavailability would be extremely detrimental in our ability to provide formation evaluation services and increase our tool fleet size allowing our company to grow. Other medium and small companies are in the same situation with a finite amount resources to pursue a pure R&D effort on alternatives. Some larger companies are looking at alternatives, but are finding the Boron Trifluoride with 1/7 the sensitivity of the Helium-3 type detectors will require increasing the activity of the Californium-252 or Americium-241 Beryllium source strengths.

## ALTERNATIVE TO HELIUM-3

The substitute for Helium-3 detectors, Boron Trifluoride (BF3), however it is much less sensitive to the thermal neutron detector as required by our industry. The majority of the sources used with neutron tools are Americium-241 Beryllium (Am-241Be), however, most recently due to Americium supplies being limited; more companies are utilizing Califorium-252 (Cf-252) in its place. Most all of these sources are in the 5 Curie (with some older 3 Curie sources used in cased hole operations) up to 20 Curies. With the decreased sensitivity of Boron Trifluoride, the strength of these neutron sources would have to be increased to achieve the statistical results needed for industry.

There are other concerns with Boron Trifluoride. The USDOT has classified this gas has a hazardous material and cannot be shipped without a US DOT special permit. Shipping by air in the US also requires classifying it as Toxic Inhalation Class 2.3. For international shipment it is restricted to Cargo Only Aircraft and classified as Toxic Inhalation Hazard Class 2.3 and Corrosive Class 8. This provides for some packaging and logistic challenges moving tools with detectors with this type of gas in the detector. Not a good solution with the mobility required for well logging tools.

## CONCLUSION

Oil and gas exploration within the U.S. is a vital part of our national security and lessens our dependence on foreign oil and gas. The shortage of Helium-3 is starting to impact our entire industry. As rig counts increase and the request for well logging increases it will require more tools to be in service ready to go. Large companies can take stock piles of tools not in service during the slowdown in the last 2 years and put them back in service. Smaller companies will have less of a stock pile of tools not in service to pull from.

With small companies such as ThruBit trying to increase our market penetration it creates an extra hardship limiting our ability to grow and bring our new technology to the market place.

Larger companies have the financial and human resources to pursue extensive research and development to look at potential alternatives in detector technologies. Smaller companies are not as fortunate - they cannot afford extensive research and development. Their commercial viability comes into question along with their ability to sustain their business. These smaller companies are also in a situation where they cannot afford the extensive research and development of looking at alternatives to their current supply of tools.

I want to personally thank you for the opportunity to discuss this important issue involving the Oil & Gas Well Services Industry today.

In: Helium-3: Critical Uses and Supply Challenges ISBN: 978-1-61470-823-0
Editors: Gabriel A. Ross 

*Chapter 9*

# Testimony of W.P. Halperin, Physics Professor at Northwestern University, before the U.S. House of Representatives, Committee on Science and Technology, Subcommittee on Investigations and Oversight, Hearing on "Caught by Surprise: Causes and Consequences of the Helium-3 Supply Crisis"

*W.P. Halperin*

Mr. Chairman and members of the committee, thank you for the opportunity to testify about the negative impact on scientific research caused by the shortage of helium--three. I am a physics professor at Northwestern University, and I rely heavily on helium--three to carry out scientific research at low temperatures and have been involved in this work since 1970. Low--temperature research is essential for studying properties of materials, such as superconductivity, and magnetism, and for developing various advanced materials. Low--temperature research is also critical to future improvements in metrology and high--speed computation, including quantum information technology. Shortages of helium--three, driven by increased homeland security demands and decreased production capability, are already creating major difficulties in these areas of research.

Let me briefly review the salient points. Helium-three is a gas and a byproduct of the radioactive decay of tritium, an essential element of nuclear weapons. Following the Second World War, as the nuclear stockpile grew, stocks of helium--three grew commensurately, reaching about 230,000 liters by the year 2000. From 2001 to the present, these stocks have been drawn down at a rate far in excess of today's global production, estimated to be approximately 20,000 liters/year. The use of helium-three as a detector of radioactive materials at ports, airports and border crossings, combined with the growth of medical, commercial and scientific applications, is responsible for the extraordinary increase in demand.

Absent new production sources, it is now impossible to serve the estimated need of 70,000 liters/year. It may be possible to find alternatives to the use of helium--three for some applications, but for others the unique physical properties of helium--three are essential.

Scientific research at low temperatures is the signature example of an area in which helium--three is irreplaceable. Without adequate supplies, such research will cease entirely. To put the importance of such research in context, I note parenthetically that twelve Nobel Laureates in physics in the past 25 years owe their accomplishments in some important measure to the availability of helium--three. Cases in which substitutes might be found for helium--three include neutron detection at facilities such as at the Spallation Neutron Source (SNS) at Oak Ridge National Laboratory, oil and gas well evaluation, building construction technology and the improvement of lasers.

The issue perhaps is best illustrated by a personal experience. In October 2008 I sought information about availability and pricing from several well--known distributors of helium--three gas. I spoke with representatives of Sigma Isotec, Cambridge Isotope Labs, Icon Isotope Services, Isoflex USA, Chemgas, and Spectra gas (now Linde Electronics and Speciality Gases) and learned that only the latter two had any supply, but their prices were extraordinarily high: $800 to $2,000/liter. It was 5 to 10 times higher than I had expected and well outside of my research budget plan.

The following summer I received more bad news. Oxford Instruments, the largest supplier of low temperature refrigerators, contacted me, to say that the company could not obtain any helium--three from their supplier, Spectra Gas. Discussions among attendees at a subsequent international low--temperature physics conference revealed that the shortage was global. Although the shortage took many of us by surprise, I later learned that some government officials had been aware of the problem for some time but had not shared this information.

In the fall of 2009, Nobel Laureates Doug Osheroff and Bob Richardson, on behalf of a low--temperature working group of which I was a member, wrote to Bill Brinkman, Director of the Department of Energy's Office of Science, to express concern about the shortage of helium--three for low temperature research. Conversations with DOE ensued, but to date requests by scientists and refrigerator companies often go unanswered or unmet. Young scientists, especially, find themselves without access to this essential resource.

Many of us are also concerned that without adequate access to helium--three, instrumentation companies may soon be forced out of business. Janis Research is an example. Janis has been guaranteed an allocation, but the helium has not been delivered and the sales interruptions place the company at risk. Should Janis and other companies stop providing refrigerators, low ---temperature science will end.

Dr. Brinkman requested that our working group assess the critical needs in low temperature science. The principal finding of our recently completed survey is the following: In a ten year interval, from 1999 to 2009, the purchase of helium--three for low temperature science averaged 3,500 liters/year and was growing at approximately 12%/year world--wide. (Survey details are posted at http://www.qfs2009.northwestern.edu/survey/ and attached to my written testimony.)

On a personal note, I have an immediate need in my laboratory for 20 liters of helium--three. Spectra Gas, the sole provider of helium--three released by the Department of Energy, has not responded in the five months since I made my request, and my National Science Foundation supported research is now in jeopardy.

In conclusion, we must recognize the diversity of needs for helium--three and adopt the following strategies: explore alternative technologies; establish effective communication among all stake holders; implement recycling and conservation; redesign critical--need instrumentation to be more efficient; and develop new sources of helium--three.

I would be pleased to answer your questions.

# SURVEY OF CRITICAL USE OF $^3$HE FOR CRYOGENIC PURPOSES

**Results of the Survey**
Northwestern University
January 21 to February 5, 2010
*You may also download the results as a PDF.*

The rare isotope of helium, $^3$He, has critical strategic importance. One of it's applications is to achieve low temperatures through refrigeration and measuring devices, mostly in the pursuit of fundamental knowledge, providing the essential building blocks for engineering and technology for our future. Cryogenic use of $^3$He is critical in that there is no alternative to reaching a range of more than 4 orders of magnitude of temperature from 1 K to as low as 10-4 K. Here basic scientific investigations require $^3$He for the study of quantum systems, including information technology, magnetism, and super-conductivity. Its recent short supply and extraordinary high price has posed serious problems for the scientific community. The purpose of this survey was to document as accurately as possible world-wide use of $^3$He in the past ten years as a framework for future cryogenic allocations and to evaluate the impact of research that uses $^3$He.

The survey is restricted to senior or principal scientific investigators, who are representatives of their respective research groups. The survey solicitation was sent to the e-mail list serves of the International Conference on Low Temperature Physics, LT25; the International Symposium on Quantum Fluids and Solids, QFS2009; a list of principal investigators using cryogenic $^3$He in their research grants from the National Science Foundation, the Program in Condensed Matter Physics; a list of principal investigators using cryogenic $^3$He in their research grants from the Department of Energy, the Program in Basic Energy Sciences. These totaled approximately 2,300 including members of the communities including students, research associates, postdoctoral fellows, scientists, and finally, the principal or senior investigators who were asked, on behalf of their groups, to respond to the survey.

This survey and a copy of the results were posted at:_http://www.qfs2009.northwestern.edu/survey/.

## Survey Results

| | |
|---|---|
| Number of senior investigator respondents: | 206 |
| USA respondents: | 98 |
| Total $^3$He purchases, yearly average over ten years: | 3,469 L/year |
| maintenance and samples gas from research groups: | 1,141 L/year |
| new instruments (mostly refrigerators) from companies: | 2,328 L/year |
| $^3$He for cryogenic purposes purchase last year (2009): | 3,828 L |
| Price of $^3$He last year, average (2009): | 930 $/L |
| Scientific programs requiring cryogenic $^3$He (fraction of total): | |
| Quantum Fluids and Solids | 8 % |
| Superconductivity | 24 % |
| Quantum Information | 7 % |
| Mesoscopic Physics | 12 % |
| Magnetism | 12 % |
| Electronic Materials | 10 % |
| Quantum Resonators | 3 % |
| Quantum Transport | 12 % |
| Refrigeration Instrumentation | 8 % |
| Detector Instrumentation | 3 % |
| Other | 2 % |
| Graduate student training using cryogenic $^3$He, graduated in ten years: | 3,349 students |
| Postdocs hired in ten years using cryogenic 3He: | 2,322 postdocs |
| Research funding in ten years requiring cryogenic 3He: | 2.65 billion $ |

*Comments on growth in the cryogenic use of $^3$He:*

Sufficient information in the responses was given to determine the following growth in requirements for cryogenic use of $^3$He. Yearly increases in purchases for cryogenic $^3$He are 12% per year on average. The increase in cost in the past three years has been approximately a factor of 4 to 5 on average.

| | |
|---|---|
| 2005 | 23 % |
| 2006 | - 1 % |
| 2007 | 30 % |
| 2008 | - 20 % |

| | |
|---|---|
| 2009 | 26 % |
| yearly average | 12 % |

*Comments on impact from research that uses cryogenic $^3$He:*

The significant impact of research that uses cryogenic $^3$He includes 335 graduate student PhD's awarded per year and 232 postdoctoral fellows hired per year (numbers adjusted for response rates determined as described in b) below averaged over the past ten years). Additionally, all theoretical work related to experimental research that uses cryogenic $^3$He would not have taken place without this range of temperature for quantum condensed systems, substantially increasing the student, staff, and funding impacts beyond that shown in this survey.

*Reporting methodology:*

a) The following nine companies provide cryogenic $^3$He instrumentation and reported their sales of $^3$He, presented above in aggregate form: Bluefors, Chase Research Cryogenics, Cryomagnetics (including Cryoconcepts), Janis Research, Lakeshore Cryotronics, Leiden Cryogenics, Oxford Instruments, Quantum Design.

b) Purchases of $^3$He, not as a part of commercial instrumentation, made by individual research groups, reported above, were adjusted by a survey response fraction of 51%. This fraction is defined by the USA pool and was assuned to be valid elsewhere in the world. The fraction is defined as the number of USA principal investigators responding to the survey divided by the total number of funded USA principal investigators identified by program mangers from the NSF/CMP and the DOE/BES. Error in corrections for survey response rate is relatively small since 2/3 of the cryogenic $^3$He is purchased by the instrumentation companies for which we have an accurate total response.

c) The responses were examined one-by-one to avoid duplication and improper submission and to be sure that each submission represented only one research group.

## Biographical Sketch

*Personal Data:* William P. Halperin
Address: Dept. Physics, Northwestern University, Evanston IL 60208
e-mail w-halperin@northwestern.edu

a) *Professional Preparation*

| | | |
|---|---|---|
| Bachelor of Science, | B.Sc.,1967 | Queen's University, Kingston, Ontario |
| Master of Science, | M.Sc., 1968 | University of Toronto, Ontario |
| Doctor of Philosophy, | Ph.D.,1975 | Cornell University, Ithaca, NY |
| Postdoctoral Fellow | (1974-75) | Cornell University, Ithaca, NY |

b) *Appointments and Honors:*

Science Advisory Board, Center of Excellence, Low Temp. and Quantum Devices, Helsinki Univ. of Tech. 2009-

International Advisory Committee, High Magnetic Field Laboratory, Chinese Academy of Sciences, Hefei 2009- Fellow, The Institute of Physics, (2004-)

Editorial Board, Journal of Low Temperature Physics, Springer, (2004-)

Chair, External Advisory Committee, National High Magnetic Field Laboratory (2004-)

Regional Editor for North America, New Journal of Physics, Inst. of Phys. (2002 -)

John Evans Professor of Physics (2001-)

Director, Interdisciplinary Science Program, Northwestern University (1999-)

Chairman, Department of Physics and Astronomy (1990-1995)

Fellow, American Physical Society (1995)

Editor, Progress in Low Temperature Physics, Elsevier (1995-)

Professor, Northwestern University (1986-present)

Associate Professor, Northwestern University (1981-86)

Alfred P. Sloan Fellow (1977-81)

Assistant Professor, Northwestern University (1975-81)

Resident Associate, Argonne National Laboratory (1979 -85)

c) *Publications (10 selected from 201):*

**Spatially Resolved Electronic Structure Inside and Outside the Vortex Core of a High Temperature Superconductor,** V. F.

Mitrovic, E. E. Sigmund, M. Eschrig, H. N. Bachman, W.P. Halperin, A.P. Reyes, P. Kuhns, W.G. Moulton, Nature **413**,501 (2001).

**Two-dimensional Superconductivity**, B. Chen, W.P. Halperin, P. Guptasarma, D.G. Hinks, V.F. Mitrovic, A.P. Reyes, and P. Kuhns, Nature Physics **3**,239 (2007).

**Discovery of an Excited Cooper Pair State in Superfluid $^3$He, J.P. Davis**. J. Pollanen, H. Choi, J.A. Sauls, and W.P. Halperin, Nature Physics, **4**,571 (2008).

**Evidence for Complex Superconducting Order Parameter Symmetry in the Low Temperature Phase of UPt3 from Josephson Interferometry**, J.D. Strand, D.J. Van Harlingen, J.B. Kycia, and W.P. Halperin, Phys. Rev. Lett., **103**, 197002 (2009)

**Magnetic Impurities in the Pnictide Superconductor** Ba1-xKxFe2As2, S. Mukhopadhyay, S. Oh, A.M. Mounce, M. Lee, W.P. Halperin, N. Ni, S.L. Bud'ko, P.C. Canfield, A.P. Reyes, and P.L. Kuhns, New J. Phys. **11**,055002 (2009).

**Quantum Size Effects in Metallic Powders**, W.P. Halperin, Rev. Mod. Phys. **58**, 533 (1986).

**Discovery of the Acoustic Faraday Effect in Superfluid $^3$He-B**, Y.Lee, T. Haard, W. P. Halperin, and J.A. Sauls, Nature, **400**,431 (1999).

**Antiferromagnetism in the Vortex Cores of** YBa2Cu3O7-d, V. F. Mitrovic, E. E. Sigmund, W.P. Halperin, A.P. Reyes, P. Kuhns, W.G. Moulton, Phys. Rev B. Rapid **67**,220503 (2003).

**Surface Specific Heat and Andreev Bound States**, H. Choi, J.P. Davis, J. Pollanen, and W.P. Halperin, Phys. Rev. Lett. **96**, 125301 (2006).

**Intrinsic Impurity in the High Temperature Superconductor $Bi_2Sr_2CaCu_2O_{8+d}$**, Bo Chen, Sutirtha Mukhopadhyay, W.P. Halperin, Prasenjit Guptasarma, and D.G. Hinks, Phys. Rev. B 77,052508 (2008).

In: Helium-3: Critical Uses and Supply Challenges ISBN: 978-1-61470-823-0
Editors: Gabriel A. Ross 

*Chapter 10*

# STATEMENT OF JASON C. WOODS, PH.D., ASSISTANT PROFESSOR RADIOLOGY, PHYSICS, AND MOLECULAR BIOPHYSICS, WASHINGTON UNIVERSITY IN ST. LOUIS, MISSOURI, BEFORE THE U.S. HOUSE OF REPRESENTATIVES, COMMITTEE ON SCIENCE AND TECHNOLOGY, SUBCOMMITTEE ON INVESTIGATIONS AND OVERSIGHT, HEARING ON "CAUGHT BY SURPRISE: CAUSES AND CONSEQUENCES OF THE HELIUM-3 SUPPLY CRISIS"

*Jason C. Woods*

## SUMMARY

If we ask seasoned pulmonologists today how much the practice of pulmonary medicine has changed in the last 25 years, responses will largely be that very little has changed—a few new drugs are available, but there is largely the same technology for measuring lung function and for treatment. $^3$He MRI, however, is beginning to emerge as a new "gold standard" and revolutionary biomarker for measuring pulmonary function and structure. Its high signal creates detailed images of lung ventilation and dynamics, and its physical properties allow precise measurement of alveolar size, microstructure, and regional lung function. This makes $^3$He MRI particularly sensitive to changes in both global and regional lung function and structure. We are at the cusp of leading pulmonary medicine to a renaissance of new drug development and image-guidance of surgical interventions for various lung diseases, such as asthma, fibrosis, and COPD, which currently affect 11% of the US population. This imaging technology, as I speak, is currently serving as a catalyst for pulmonology to see significant advances in the next 10 years. A lack of supply of $^3$He gas will stifle these advances.

This $^3$He shortage affects *my* research acutely; it affects my employees and collaborators, and the research and livelihood of MRI groups in at least 11 US universities and at least that many universities abroad. For me personally, a lack of gas will likely mean that my research is shut down, and I would join the ranks of the unemployed. To be clear, however, I think the larger impact of this technology is not on my research group but in drug development, in much easier determination of the effectiveness of new pharmacologic agents, and in guiding new minimally-invasive interventions (my most recent work). The lack of big leaps forward in drugs to treat lung diseases—asthma, COPD, pulmonary fibrosis—has largely been due to the combination of the exceptional cost to bring drugs to market and the lack of a precise biomarker to determine changes in the lung. Pulmonary function tests, the decades-old standard in pulmonary medicine, have notoriously high measurement errors. Obstructive lung diseases (asthma and COPD), taken together, afflict approximately 35 million Americans; COPD alone is the 4$^{th}$ leading cause of death and is the only major cause of death that is steadily increasing [1, 2]. The financial and human impacts of the shortage are significant.

One recent example of drug efficacy testing illustrates the lack of a precise biomarker and its impact: in 2007 GlaxoSmithKline released results of an Advair study, entitled "Toward a revolution in COPD health (TORCH)." The total cost was estimated at $500 million dollars for this study in over 6,000 moderate and severe COPD patients. The study endpoint was death from all causes, which ranged from a high of 16% to a low of 12.6% for those on Advair. The key question was "Does Advair reduce mortality by as much as 20%?" Unfortunately for GSK, the question remained unanswered, because the statistical p-value of the difference was 0.052. This means the difference in mortality had a 5.2% chance of occurring randomly, whereas the generally accepted limit is 5%. This $500M thus was largely wasted; the company couldn't answer the question about benefit, and patients and society received no benefit or increased understanding from the study. If the $^{3}$He diffusion MRI techniques that our group has developed, for example, were used as a biomarker and endpoint (not possible when the study began), 6,000 patients could have turned into fewer than 500 patients, saving around 90% of the cost of the study, or $450M. And the question about efficacy would likely have been answered. This is only one example of the type of significant impact that I think $^{3}$He MRI is going to have on pulmonary medicine.

There has been some discussion in the scientific literature about using hyperpolarized $^{129}$Xe instead of $^{3}$He gas for specific future studies, and for some studies this *may* be a viable alternative within the next 5-10 years [3], though the intrinsic physical properties of $^{129}$Xe reduce the signal by a factor or 3-5 compared to $^{3}$He. Some damage to the field could be tempered by outside assistance in developing this infrastructure and technology. However, many studies, like my NIH-funded research, rely upon $^{3}$He's large diffusion coefficient for large- distance measurements, and for this xenon will *not* be an alternative [4]. On the bright side, the $^{3}$He that we use is nearly 100% recyclable, but we do not yet have the recycling technology in place to begin to do this. I believe firmly that the development of efficient and commercially viable recycling schemes will allow this important work to continue, with a total allotment of around 2,000 Liters STP per year.

Lastly, I note that in 2009 an allocation of $^{3}$He was made specifically for the NIH-funded medical imaging community. This was offered through Spectra Gases (now Linde Gas) at $600/L STP – an approximately 500% increase over previous years. Because the price of $^{3}$He increased so quickly and by so much, research groups (who have strict budgets from federal or private grants) were not able to plan for the cost increase and are now scrambling for supplementary funding sources. This is the reason why all of the $^{3}$He recently set aside for various medical imaging groups has not been instantly purchased.

## BACKGROUND

Conventional MRI relies upon a large magnetic field to generate a net alignment of nuclear spins (generally within the hydrogen atoms of water molecules), which can be manipulated to create images with high contrast. The technology allows images to answer specific questions about structure and function of the brain, joints, or other parts of the body [5, 6]. MRI of gas is not generally used, since the density of a gas is about 1000 times less than tissues, and there is not enough signal to generate an image. The unique properties of the $^{3}$He atom allow us to align a large fraction of its nuclear spins via a laser polarization technique with a magnetic field; this is often called "hyperpolarization" [7, 8]. Hyperpolarized $^{3}$He gas has signals enhanced by a factor of 100,000 or more—allowing detailed images of the gas itself to be generated in an MRI scanner. Since helium gas (either $^{4}$He or $^{3}$He) has a solubility of essentially zero and is arguably the most inert substance in the universe, inhaled hyperpolarized $^{3}$He allows the generation of exceptional quality, gas-MR images of ventilated lung airspaces with no ionizing radiation or radioactivity [9]. Further, traditional technologies for measuring pulmonary function (*e.g.*, pulmonary function tests or nuclear ventilation scans) have either high errors on reproducibility or low content of regional information. While x-ray CT has some potential for quantifying lung structure (not function), its large amount of ionizing radiation raises cancer risks and prevents it from being used in longitudinal studies for drug development or in vulnerable populations, such as children [10, 11]. $^{3}$He is inert and has proven to be very safe in studies to date (helium-oxygen mixtures[12] are used routinely in pulmonary and critical care); it is, however, currently regulated as an investigational drug by the US FDA.

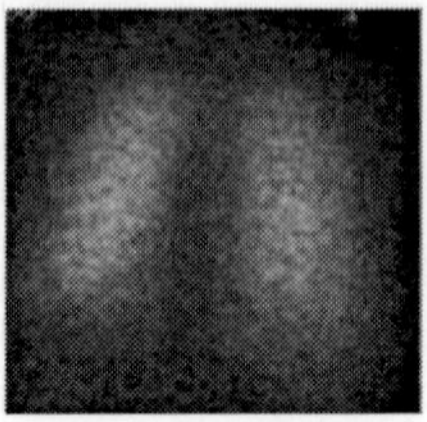

Figure 1. Nuclear ventilation scan.

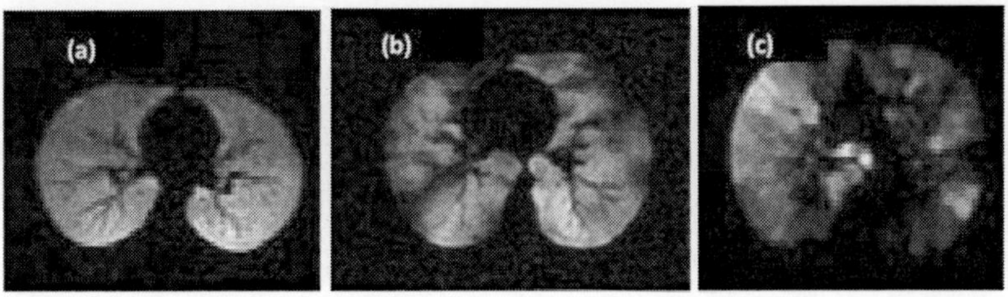


Figure 2. High-resolution transverse ventilation images of (a) a healthy volunteer and patients with (b) asthma and (c) COPD. This type of ventilation imaging represented a

huge leap forward in our ability to visualize and quantify regional ventilation distribution and dynamics.

# THE REVOLUTION OF $^{3}$HE MRI ON PULMONARY IMAGING

## Ventilation

Previous technologies for imaging pulmonary ventilation generally involved the inhalation of radioactive gas over a period of one to several minutes, and then detecting what parts of the chest emitted the most radioactivity over several minutes. This technology (nuclear ventilation scans) had low spatial and temporal resolution (Figure 1). $^{3}$He ventilation MRI represented a clear step forward in depicting not only precise, 3-D regional ventilation, but also in beginning to understand the regional dynamics of human ventilation in health and disease.

At present, $^{3}$He ventilation imaging is being used in a wide variety of studies and holds high promise in increasing our understanding of the regional effects of asthma and its treatment [13-16], in addition to COPD, and various types of lung fibrosis [17, 18]. For example, it was recently found (Figure 2) that many ventilation defects persisted over time, opening the door to new regional treatments for asthma—an idea not previously pursued [19]. Because asthma is the most prevalent pulmonary disease in the US, improved medical and interventional therapies, facilitated by $^{3}$He MRI, can significantly improve care and lower health care costs.

## Diffusion and in-Vivo Morphometry

Three unique physical properties of $^{3}$He make it particularly well suited for measuring lung airspace size, geometry, and connectivity, by quantifying its restriction to thermal diffusion in the lung. These properties are 1) its small size (and thus large thermal diffusion coefficient), 2) its lack of solubility in tissue, and 3) its long relaxation time, T1. Since $^{3}$He is insoluble and has a large diffusion coefficient, collisions with airway and alveolar walls restrict the movement of the gas. This restriction can be measured and quantified using diffusion MRI. In fact, our group in particular has had a focus on $^3$He diffusion MRI; we have shown that the technique is extraordinarily sensitive to airspace enlargement and has better discrimination than quantitative histology—the gold standard for airspace quantification in lung parenchyma (Figure 3). We have recently shown that the technique can be used to measure the size and geometry of alveolar ducts—allowing regional morphometry of the human lung, in vivo (Figure 4). These types of measurements are not available by any other noninvasive technique and represent a leap forward in our understanding of lung microstructure and our ability to quantify early disease.

Airspace enlargement (emphysema) is a significant component of chronic obstructive pulmonary disease (COPD)—the only leading cause of mortality with dramatic *increases* in the US and the world [2]. Quantifying this airspace enlargement in a reliable and precise way, as $^3$He MRI easily can, has enormous potential therapeutic benefit for patients with COPD. No other measurement modality has such potential to detect early disease, disease progression, or to quantify microstructural parameters in the $^3$He MRI can. Figure 4 demonstrates this in two volunteers with normal lung function by pulmonary function test and with normal CT scans; $^3$He MRI, however, can distinguish early lung disease in the smoker at right. This extraordinary sensitivity to early disease makes it a prime biomarker for use in drug development and efficacy testing.

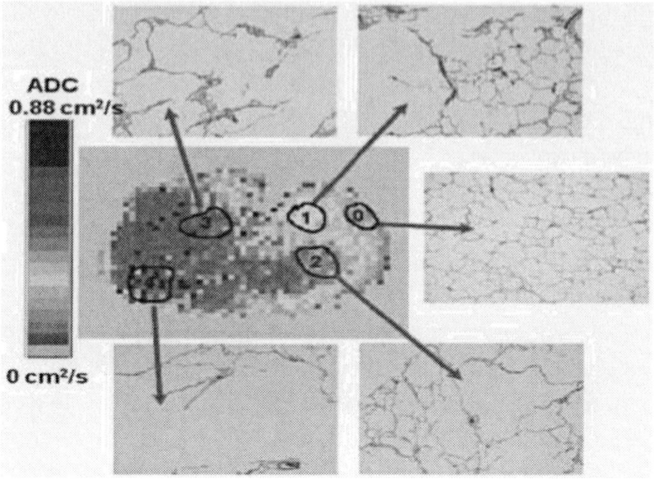


Figure 3. $^3$He diffusion image (color) of a human lung with heterogeneous COPD, demonstrating the ability of the $^3$He ADC to predict morphometry.

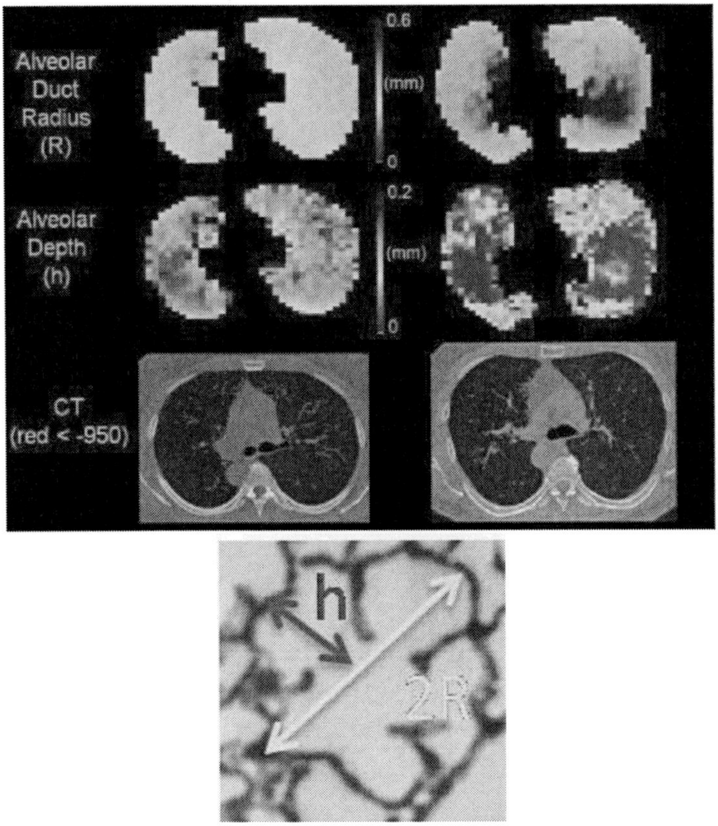


Figure 4: Maps of lung microstructure (color) obtained from gas diffusion imaging, in two volunteers with normal lung function and normal CT, yet the patient on the right is a current smoker and has a greatly reduced alveolar depth. *R* and *h* are depicted below within an individual alveolar duct (perpendicular to the page).

One particularly unique quality of $^3$He comes from the combination of its large gas diffusion coefficient and insolubility in tissue. This allows us to the diffusion of the gas over very long distances (2-5 cm) and has been called "long-range diffusion". Because these distances are larger than any acinar dimension, the technique is sensitive to the extent of "collateral" or short-circuits pathways other than the airway tree in the lung. These collateral pathways are essential to quantify for two minimally-invasive interventions that are being developed for end-stage COPD: transbronchial stents (Broncus Technologies, Inc.; Mountain View, CA) and one-way exit valves in segmental bronchi (Spiration, Inc.; Redmond, WA). My most recent NIH-funded research involves the use of long-range $^3$He diffusion to guide and predict the efficacy of these minimally-invasive interventions under development. Early results are quite promising, and demonstrate that the imaging will do quite well at guiding the therapy, but the shortage of $^3$He has had a negative impact on the study.

## Regional Pulmonary Oxygen Monitoring

The long relaxation time $T_1$ of $^3$He and its sensitive dependence on oxygen concentration allow us to measure the regional partial pressure of oxygen in the lung. Maps of this partial pressure ($pAO_2$) in the lung can be used to understand regional pulmonary blood flow and diffusion of oxygen into capillaries--the essential purpose of the organ. Not only can $pAO_2$ be used to measure deficiencies in the partial pressure of oxygen, but it can be employed to understand the regional relationship between structure and function in the lung, at its most fundamental level (oxygen and $CO_2$ transfer). Again, this is a technique only possible via $^3$He MRI.

## Partial List of Currently Funded $^3$He Imaging Projects in North America

The following list of current $^3$He MRI research projects is far from complete but represents the broad range of lung diseases studied and research funded by both the NIH and by US-based private industry:

***Assessing drugs for treatment of cystic fibrosis****:* University of Massachusetts (Dr. Albert, *et al.)*

***Detecting early and preclinical COPD****:* Washington University (Dr. Yablonskiy, *et al.*)

***Detection of pulmonary metastases with*** $^3$***He****:* Duke University (Dr. Driehuys, *et al.*)

***Detecting and treating pulmonary embolism:*** University of Massachusetts (Dr. Albert, *et al.*)

***Diffusion kurtosis imaging in asthma, COPD and in the lungs of 9/11 NYC firefighters***: New York University (Dr. Johnson, *et al.)*

***Drug Efficacy in preclinical models of asthma and COPD***: Duke University (Dr Driehuys, *et al.*)

***Early detection of bronchiolitis obliterans syndrome in lung transplant recipients***: Washington University (Dr. Woods, *et al.*)

***Evaluation of endobronchial interventions for COPD***: Washington University (Dr. Woods, *et al.*), Robarts Imaging Institute (Dr. Parraga, *et al.*, University of Virginia (Dr. Altes, *et al.)*

***Evaluation of a novel treatment for asthma:*** University of Virginia (Dr. Altes, *et al.)*

***Evaluation of a novel treatment for cystic fibrosis***: University of Virginia (Dr. Mugler, *et al.)*

***Imaging of small-animal models of diseases***: Duke University (Dr. Johnson *et al.),* Washington University (Dr. Woods, *et al.)*

***In-vivo morphometry with $^{3}$He diffusion MRI***: Washington University (Dr. Yablonskiy, *et al.*)

***Measuring regional pulmonary oxygen pressure by $^{3}$He MRI:*** University of Pennsylvania (Dr. Rizi, *et al*)

***Monitoring Progression of COPD***: Duke University (Dr Driehuys, et al.), Robarts Imaging Institute (Dr. Parraga, *et al.*), University of Virginia (Dr. Mugler, *et al.*), Washington University (Dr. Yablonskiy, *et al.)*

***Neonatal ventilation and dynamics under mechanical ventilation***: Harvard University (Dr. Patz *et al.*), University of Virginia (Dr. Miller, *et al.)*

***Noninvasive methods for measuring alveolar surface area***: Harvard University (Dr. Patz, *et al.*), Washington University (Dr. Yablonskiy, *et al.*)

***Persistence of Ventilation Defects in patients with asthma:*** University of Virginia (Dr. Altes, *et al.*), University of Massachusetts (Dr. Albert, *et al.*)

***Predicting ventilation changes caused by radiation therapy:*** Robarts Imaging Institute (Dr. Parraga, *et al.*), University of Virginia (Dr. Mugler, *et al.)*

***Probing the fundamental limits of MRI resolution by diffusion:*** Duke University (Dr. Johnson, *et al.*)

***Pulmonary Gas flow Measurements and Dynamic $^{3}$He MRI of the Lungs:*** New York University (Dr. Johnson, *et al.)*

***A Specialized Clinically Oriented Center of Research for COPD:*** (Dr. Holtzman and Dr. Woods, *et al.)*

# THE $^{3}$HE SHORTAGE AND ITS EFFECTS

## Timeline

Late in 2008 our research group and others became aware that there was a supply issue with $^{3}$He gas, through conversations with Spectra Gases, Inc. We immediately purchased some gas to continue imaging studies in COPD patients. In March, 2009, we were told there was no gas available for medical applications and that the price of non-medical $^{3}$He had risen to near \$400/L STP. Conversations with colleagues at the University of Virginia, Harvard University, and the University of Pennsylvania confirmed that others were also unable to purchase $^{3}$He gas. In April - June of 2009, we worked with Spectra Gases and other universities to state our $^{3}$He requirements to continue NIH- and NSF-funded research in 2009; Spectra Gases then met with the Department of Energy (DOE) in July and August to make clear that US Government-funded research was being affected. In August 2009, Spectra approached me and the other officers of the Hyperpolarized Media Study Group of the International Society for Magnetic Resonance in Medicine (the primary professional organization for $^{3}$He MRI researchers) to write a letter to the Isotope Work Group of the DOE, stating how $^{3}$He is unique in medical imaging and that a significant amount of NIH-funded research would be effectively shut down without access to the small amount of gas that our community uses (2000 L STP / year, approximately). Dr. William Hersman and I drafted this letter, dated September 4, 2009; it is attached to the end of this written testimony. In October 2009 we were notified by Spectra Gases that an algorithm for obtaining a small amount of $^{3}$He gas for NIH-funded studies had been achieved. In order to obtain any gas, we were to list each federally-funded grant's title and number, and for each a requested amount of gas for the subsequent 6 months of usage. $^{3}$He was offered to our group for $600/ L STP, an approximately 500% increase from previous years. I also drafted a letter in support of Spectra's modification of their permit for $^{3}$H (tritium) limits with $^{3}$He, in addition to letters of support for allocation of $^{3}$He to two non-US researchers who do important work; these are also attached to this testimony. At a recent AAAS meeting (April, 2010), it was made clear that the White House Office of Science and Technology Policy (OSTP) had been diligently and actively pursuing a solution to this shortage by facilitating discussions between DOE and DHS. My understanding is that OSTP was helpful in (perhaps in large part responsible for) the 2009 and 2010 allocation of $^{3}$He gas to NIH-funded projects.

## Impact of the Shortage upon Medical Imaging Research

While I have stated that I think the biggest impact of $^{3}$He MRI *technology* is in drug development, efficacy monitoring, and in guiding new minimally-invasive interventions, the impact of the shortage was most keenly felt by those of us in the middle of performing NIH- and industry-funded research studies. Some of us (like our group at Washington University) were able to continue to perform studies at a lower rate and were able to purchase gas at $600/L STP, once it became available. Other groups, such as the Robarts Imaging Institute, have not been able to continue $^{3}$He studies, even if these studies were funded by US companies. Even for US, NIH-funded researchers, however, the price of $^{3}$He increased so quickly and by so much that research groups were not able to plan for the cost increase and are now scrambling for supplementary funding sources. This is the reason why all of the $^{3}$He recently set aside for various medical imaging groups has not been instantly purchased. The shortage has had a significant negative impact on the continued productivity of our research community and on the probability of future research. Importantly, if sufficient $^{3}$He is not allocated to medical imaging at reasonable cost, this will likely curtail the revolution in pulmonary medicine currently in progress.

## Financial and Scientific Impact

It is difficult to gauge the precise financial impact of the $^3$He shortage on the field of hyperpolarized-gas MRI. It is clear that fewer studies are being conducted and planned as a result of this shortage. It is probably safe to say that all studies mentioned previously have been scaled back by a factor of 2 or more. By my count, the National Institutes of Health are currently supporting at least 25 active projects requiring $^3$He, with over $4M allocated for FY2010. If we assume similar funding for the past 8 years, with less funding before that, this represents an investment of over $32M via NIH funding alone. When added to the significant (but more difficult to quantify) investment from the NSF, private and public universities, and private industry, the total investment in $^3$He MRI is likely between $60M and $100M over the past 10 years.

While the above numbers represent an enormous investment in $^3$He polarization and MRI infrastructure, it is my opinion that the biggest financial impact of the shortage is on future drug development, efficacy monitoring, and in guiding new surgical and minimally-invasive interventions. Through the use of more precise biomarkers, such as we have developed via $^3$He MRI, the number of patients required to determine the true efficacy of a drug or device can be reduced by large fractions (up to 90% by a recent calculation from our techniques), which would translate directly into proportionate cost savings. The GSK example of the TORCH study mentioned in the Summary is illustrative. The key question was "Does Advair reduce mortality by as much as 20%?" Unfortunately for GSK, the question remained unanswered after studying 6000 patients and expending $500M, because the statistical significance was not high enough to determine an answer to the vital question. If the $^3$He diffusion MRI techniques that have been discussed here were used as a biomarker and endpoint (not possible when the study began), 6,000 patients could have turned into fewer than 500 patients, saving around 90% of the cost of the study, or $450M. The question about efficacy would likely have been answered, and the company could have devoted its efforts to the marketplace, if successful, or to newer and more innovative solutions, if unsuccessful.

The scientific impact of the shortage is serious. Scientific studies and investigations into lung physiology and pathophysiology and new treatments are being scaled back; without a clear solution in place, the revolution in pulmonary medicine will be at least partially curtailed. In one case that I'm very familiar with, research has ceased entirely because of a lack of $^3$He gas. The Robarts Research Institute in London, Canada was established in part with capital funding provided by and research partnerships with Merck Research Laboratories (Imaging, Westpoint PA USA) and General Electric Health Care (GEHC, Milwaukee WI). They have been performing $^{3}$He MRI studies in animal models of respiratory disease, in healthy volunteers, and patients with lung disease (COPD, asthma, cystic fibrosis, radiation-induced lung injury). Their human studies are funded by Merck, GEHC, the Canadian Lung Association and Canadian Institutes of Health Research. Without a small allocation of $^{3}$He to this institution, their entire pulmonary MRI operation will be shut down, and further investment by US companies will be lost.

## Potential Alternatives to $^{3}$He

Two noble gas isotopes ($^{3}$He and $^{129}$Xe) were originally identified as having potential for use in pulmonary MRI, since they could be hyperpolarized to 10% or more with sufficient laser power (originally very expensive and technically complex). Other gases (e.g. $^{83}$Kr, $^{21}$Ne) have potential for low levels of hyperpolarization, but their nuclear and physical properties will prevent high polarizations in bulk gas or their widespread use in human MRI. When high-power, low-cost diode laser technology became available in the 1990s, these lasers were used to produce macroscopic quantities of $^{3}$He at high polarization (~50-60%), and $^{129}$Xe at much lower polarization ($\leq$ 10%). The comparative physical properties of the gases and early hyperpolarization technology led to near-universal adoption of $^{3}$He as the gas of necessity for pulmonary gas MRI. These properties are outlined below.

1. The magnetic moment of $^{129}$Xe is only about 1/3 that of $^{3}$He; this is directly related to the signal strength in MRI. Further, the natural abundance of $^{129}$Xe is only 26%; both of these reduce the available signal in the hyperpolarized gas intrinsically by a factor of 6. Enrichment of the isotope (at significant cost, since $^{129}$Xe is close in weight to the abundant isotopes of Xe) can reduce this intrinsic signal reduction to a factor of 3 below $^{3}$He. The achievable polarization with xenon has also been historically lower than with $^{3}$He, and the delivered dose of xenon gas is limited by its anesthetic activity. In short, hyperpolarized xenon does not yield the high signal-to-noise that $^{3}$He does, which means that xenon delivers poorer quality images and less physiological information. The sum of the effects of lower magnetic moment (gyromagnetic ratio), lower abundance, lower

polarization, and lower dose add up to an approximate reduction in signal by a factor of 50. The efforts of Dr. William Hersman (XeMed, LLC) have helped to increase $^{129}$Xe polarizations, but this new technology requires new, significant capital investment by each hyperpolarized group wishing to switch to $^{129}$Xe. Even with "perfect" new technology which achieves comparable polarization and with isotopically enriched gas, the signal reduction is still intrinsically limited by the magnetic moment and limited dose—a factor of 3-5—and many experiments and clinical trials are not possible with $^{129}$Xe. This is particularly true for measurements of lung morphometry and connectivity.

2. The free diffusivity of $^3$He is extremely large, because of its low mass and small collisional cross-section. This property is crucial to measurements of long-range diffusivity in lungs, which have been shown to be more sensitive to emphysema than short-range diffusivity. By comparison, the much lower free diffusivity of xenon greatly reduces the distances that can be explored with the long-range technique. To our knowledge, no one has even reported long-range diffusion measurements in lungs with hyperpolarized xenon for this reason. Several of our NIH-funded projects rely upon a measurement of long-range $^3$He diffusion and would not be completed without the $^3$He isotope. Further, larger field gradients are required even for short-range diffusion experiments; this may require further capital costs.
3. The long T1 of $^3$He allows it to be shipped by air freight. This has been demonstrated in Europe and the Mayo Clinic (in addition to a current proposal by Dr. Hoffman's group at the University of Iowa) as a feasible business-model for polarized gas use in hospitals, removing the necessity of each hospital having its own dedicated polarizer (a requirement that has so far limited the clinical utilization of polarized gas). By comparison, the T1 of xenon is shorter (of order 2 hours), making air shipment virtually impossible to orchestrate.

$^3$He will remain a necessity for MRI researchers because of the physical properties mentioned above (specifically its high diffusion coefficient). The intrinsic properties of $^{129}$Xe will necessarily limit the images to have a factor of 3 reduction in signal compared to $^3$He images. The polarization of $^{129}$Xe has seen significant improvement in the past 3-4 years, however, and some recent images of ventilation have had acceptable contrast, even though the signals were not as high as for $^3$He. And while the relatively large solubility in tissue has an anesthetic effect on animals and humans, this property can be capitalized upon in an attempt to quantify diffusion across gas-tissue barriers. There is thus a potential role for $^{129}Xe$ in perhaps half of the future hyperpolarized-gas MRI studies.

## Recycling $^{3}$He

Since helium is not soluble in the tissues of the body, it can be very highly recoverable, yet most research groups do not have systems currently in place to recapture and compress exhaled gas. The hyperpolarized helium research community has demonstrated in the past that inexpensive technologies can be assembled for easily solvable problems within the field, and the technology for recycling of $^{3}He$ is straightforward. (For example, since $^{3}He$ is a liquid at 4 K [4 degrees above absolute zero], all other gases, particulate and biological matter can be frozen out by passing through a liquid $^{4}He$ bath at 4 K.) Both Washington University (Dr. Woods, et al.) and the University of Virginia (Dr. Miller, et al.) are currently collaborating with Walter Whitlock, of Conservation Design Services, Inc., in North Carolina, to develop commercially-viable recycling for wide use in the $^{3}He$ MRI community. This recycling collaboration is not yet funded but is currently underway. I believe that the important and significant scientific research outlined in this testimony can be sustained and performed with around 2,000 total STP liters of $^{3}He$ per year, after development of good recovery/recycling systems for $^{3}He$.

## References


[1] Bed na rek, M., et al. (2008). Prevalence, severity and underdiagnosis of COPD in the primary care setting. *Thorax, 63(5)*, p. 402-7.

[2] Mannino, D. M. & Buist, A. S. (2007). Global burden of COPD: risk factors, prevalence, and future trends. *Lancet, 370(9589)*, p. 765-73.

[3] Hersman, F. W., et al. (2008). Large production system for hyperpolarized 129Xe for human lung imaging studies. *Acad Radiol, 15(6)*, p. 683-92.

[4] Woods, J. C., et al. (2005). Long-range diffusion of hyperpolarized $^{3}He$ in explanted normal and emphysematous human lungs via magnetization tagging. *J Appl Physiol, 99(5)*, p. 1992-7.

[5] Callaghan, P. T., *Principles of nuclear magnetic resonance microscopy*. 1991, Oxford [England]; New York: Clarendon Press; Oxford University Press. xvii, 492 p.

[6] Talagala, S. L. & Lowe, I. J. (1991). Introduction to magnetic resonance imaging. *Concepts Magn Reson, 3*, p. 145-159.

[7] Gamblin, R. L. & Carver, T. R. (1965). Polarization and relaxation processes in 3He gas. *Phys Rev, 138*, p. 946-960.

[8] Walker, T. G. & Happer, W. (1997). Spin exchange optical pumping of noble-gas nuclei. *Rev Mod Phys, 69*, p. 629-642.

[9] Lutey, B. A., et al. (2008). Hyperpolarized $^3$He MR imaging: physiologic monitoring observations and safety considerations in 100 consecutive subjects. *Radiology, 248(2)*, p. 655-61.

[10] Berrington de González, A., et al. (2009). Projected Cancer Risks from Computed Tomographic Scans Performed in the United States in 2007. *Arch Intern Med, 169(22)*, p. 2078- 2086.

[11] Smith-Bindman, R., et al. (2009). Radiation Dose Associated With Common Computed Tomography Examinations and the Associated Lifetime Attributable Risk of Cancer. *Arch Intern Med, 169(22)*, p. 2078-2086.

[12] Frazier, M. D. & Cheifetz, I. M. The role of heliox in paediatric respiratory disease. *Paediatr Respir Rev, 11(1)*, p. 46-53; quiz 53.

[13] Wang, C., et al. (2008). Assessment of the lung microstructure in patients with asthma using hyperpolarized $^3$He diffusion MRI at two time scales: comparison with healthy subjects and patients with COPD. *J Magn Reson Imaging, 28(1)*, p. 80-8.

[14] de Lange, E. E., et al. (2006). Evaluation of asthma with hyperpolarized helium-3 MRI: correlation with clinical severity and spirometry. *Chest, 130(4)*, p. 1055-62.

[15] Fain, S. B., et al. (2008). Evaluation of structure-function relationships in asthma using multidetector CT and hyperpolarized He-3 MRI. *Acad Radiol, 15(6)*, p. 753-62.

[16] Altes, T. A. & de Lange, E. E. (2003). Applications of hyperpolarized helium-3 gas magnetic resonance imaging in pediatric lung disease. *Top Magn Reson Imaging, 14(3)*, p. 231- 6.

[17] Woods, J. C., et al. (2006). Hyperpolarized 3He diffusion MRI and histology in pulmonary emphysema. *Magn Reson Med, 56(6)*, p. 1293-300.

[18] van Beek, E. J., et al. (2007). Assessment of lung disease in children with cystic fibrosis using hyperpolarized 3-Helium MRI: comparison

with Shwachman score, Chrispin-Norman score and spirometry. *Eur Radiol, 17(4)*, p. 1018-24.

[19] de Lange, E. E., et al. (2009). Changes in regional airflow obstruction over time in the lungs of patients with asthma: evaluation with 3He MR imaging. *Radiology, 250(2)*, p. 567-75.

Jason C. Woods, Ph.D.
Assistant Professor of Radiology, Physics, & Molecular Biophysics
Assistant Dean of Arts & Sciences
Director, MARC uSTAR Program

Division of Nuclear Materials Safety
U.S. Nuclear Regulatory Commission

Cc: Spectra Gases, Inc. (Jack Faught, Keith Darabos)
3434 Route 22 West
Branchburg, NJ 08876

September 4, 2009

RE: $^3$He is unique and irreplaceable - Letter to the Isotope Work Group of the US Government in relation to release of $^3$He for medical imaging.

Dear Keith,

Our understanding is that the US government's "isotope work group" may benefit from some additional information and a statement from the hyperpolarized-gas community concerning the necessity of $^3$He as the hyperpolarized isotope of choice for our pulmonary imaging studies and clinical trials. $^3$He has emerged as a new and unique standard for pulmonary imaging, both for its high signal, physical properties (specifically, a high diffusion coefficient, allowing morphometric measurements of alveolar spaces), developed infrastructure, and mature polarization technology. There has been some discussion in the literature about using hyperpolarized xenon instead of $^3$He gas for specific future studies, but it is clear to us that hyperpolarized xenon is simply not suitable to replace $^3$He at this point, both for scientific and practical reasons. We outline these reasons below. In the long run, we believe that significant scientific research can be performed with around 1000-2000 STP liters of $^{3}$He after development of good recovery/recycling systems for $^{3}$He.

1. The magnetic moment of $^{129}$Xe (the spin 1/2 stable isotope) is only about 1/3 that of $^{3}$He, and the natural abundance of $^{129}$Xe is only 26%; both of these reduce the available signal in the hyperpolarized gas. The achievable polarization with xenon has also been historically lower than with $^{3}$He, and the delivered dose of xenon gas is limited by its anesthetic activity. In short, hyperpolarized xenon does not yield the high signal-to-noise that $^{3}$He does at this point in time, which means that xenon delivers poorer quality images and less physiological information. The sum of the effects of lower gyromagnetic ratio, lower abundance, lower polarization, and lower dose add up to an approximate reduction in signal by a factor of 50. In the future, the achievable polarizations of xenon are expected to improve, though we do not view this as a viable replacement for $^{3}$He for the reasons stated herein. The efforts of Dr. Bill Hersman (XeMed, LLC) have helped to increase $^{129}$Xe polarizations, but this new technology requires new, significant capital investment by each hyperpolarized group wishing to switch to $^{129}$Xe. Even with "perfect" new technology which achieves comparable polarization and with isotopically enriched gas, the signal reduction is still limited by the gyromagnetic ratio and limited dose—a factor of 3-5—and many experiments and clinical trials are not possible with $^{129}$Xe.
2. Different physical properties of Xe and He lead to differences in the investigations that can be performed with the two gases. One big difference is the self diffusion coefficient; the free diffusivity of $^{3}$He is extremely large, because of its low mass and small collisional cross-section. This property is crucial to measurements of long-range diffusivity in lungs, which have been shown to be more sensitive to emphysema than short-range diffusivity. By comparison, the much lower free diffusivity of xenon (neat, or mixed with air or nitrogen) greatly reduces the distances that can be explored with the long-range technique. To our knowledge, no one has even reported long-range diffusion measurements in lungs with hyperpolarized xenon for this reason. Several of our NIH-funded projects rely upon a measurement of long-range $^{3}$He diffusion and would not be completed without the $^{3}$He isotope.

3. The long T1 of $^{3}$He allows it to be shipped by air freight. This has been demonstrated in Europe and the Mayo Clinic (in addition to a current proposal by the Iowa group) as a feasible business-model for polarized gas use in hospitals, removing the necessity of each hospital having its own dedicated polarizer (a requirement that has so far limited the clinical utilization of polarized gas). By comparison, the T1 of xenon is shorter (of order 2 hours), making air shipment virtually impossible to orchestrate. Further, the hyperpolarized helium research community has demonstrated that inexpensive technologies can be assembled for recapturing and recycling of the $^{3}$He gas used for medical imaging. Since helium is not soluble in the tissues of the body, it can be very highly recoverable.
4. Many groups in the field using hyperpolarized $^{3}$He have existing, funded grants, most of which are from the National Institutes of Health, representing a $100 million investment over the past decade in infrastructure and expertise. A substantial decrease in availability or increase in price of $^{3}$He would be an insurmountable burden on these groups and their research efforts. In this regard, we ask that you note that switching from $^{3}$He to xenon (assuming xenon is suitable for the proposed measurements; see above) would entail large equipment expenses, which current funding would not cover. A conversion of existing polarizers on loan from General Electric (or built in-house by scientists) would require major, expensive changes; imaging would have to occur with new rf coils (now often complex multi-receiver "phased arrays"), pulse and sequence development, and an established multi-year safety record. Even with such a large capital and time investment for xenon substitution, resultant images with $^{129}$Xe would be inferior to what our current $^{3}$He studies require.

In summary, we see a huge loss of scientific productivity, a wasting of a large investment in medical research infrastructure, a forestalling of medical advances for the US patient population, and irreparable damage to careers of scientists and students without an immediate release of $^{3}$He for use by the medical research community. While hyperpolarized xenon may, in the long term, meet some of the medical needs presently served by $^{3}$He, we do not see this making an impact without a large investment in time (several years) and research dollars (>$ 1 0M), the funds for which do not exist. Even if that investment were made, clinical trials currently underway would have to be restarted. If, on the other hand, a programmatic release of $^{3}$He over the next several years were allowed, important research with the existing infrastructure could continue. Market forces or quantity restrictions will rapidly result in technological and methodological ways of maximizing the progress using as little $^{3}$He gas as possible.

On behalf of the Hyperpolarized Media Study Group of the International Society of Magnetic Resonance in Medicine,

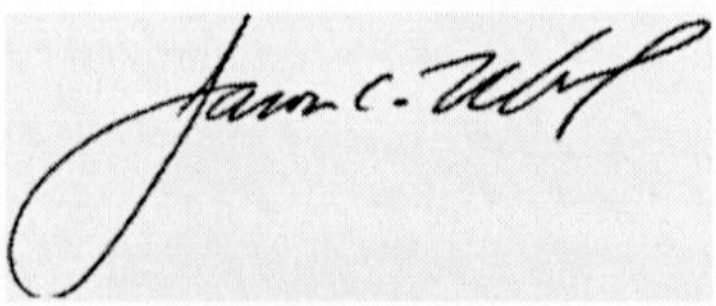

Jason C. Woods, Ph.D.
Program Chair, Hyperpolarized Media Study Group
Assistant Professor of Radiology, Physics, & Molecular Biophysics
Assistant Dean of Arts & Sciences
Washington University

F. William Hersman, Ph.D.
President, Hyperpolarized Media Study Group
Professor of Physics, University of New Hampshire
CEO, Xemed, LLC

Dr Jason C. Woods
Assistant Professor of Radiology, Physics, & Molecular Biophysics
Assistant Dean of Arts & Sciences
Director, MARC uSTAR Program

Division of Nuclear Materials Safety
U.S. Nuclear Regulatory Commission

Cc: Spectra Gases, Inc. (Jack Faught, Keith Darabos, Michael Baselice)
3434 Route 22 West
Branchburg, NJ 08876

September 30, 2009

RE: modification of Spectra Gases' permit (license # 29-30779-0 1) for tritium contained within helium gas.

To whom it may concern:

We understand that the US Nuclear Regulatory Commission may benefit from some additional information and a statement from the hyperpolarized-gas community regarding the modification of Spectra Gases' license for tritium limits within the stable isotope $^3$He, which our community uses routinely for pulmonary medical imaging via MRI. Our position is that the levels of allowed tritium within inhaled $^3$He gas (5 x $10^{-6}$ μCi/cc) is many orders of magnitude below any level potentially harmful to humans. This position is based upon our own calculations and those provided by the Department of Energy, within the Handbook, *Tritium Handling and Safe Storage* (DOE-HDBK-1 129-99). We urge that the requested modification to the license, which will allow more $^3$He to be used for NIH-sponsored and other research projects within our scientific community, be granted expeditiously.

We offer a calculation below to illustrate the safety of the allowed level of tritium within $^3$He gas. Since each research group uses a slightly different amount of $^3$He per experiment (which range from 300 mL to 1000 mL) we assume the maximum of 1000 mL $^3$He per experiment and 3000 mL $^3$He per imaging session in the calculations.

The DOE Handbook *Tritium Handling and Safe Storage* correctly notes that tritium is not readily absorbed from inhaled gases or through the skin. (This is due to its very low solubility in tissue and blood—near the low value of $^3$He gas itself.) They state on p. 5 that, "A very small fraction of the inhaled hydrogen isotopes, in gaseous form, is not exhaled, but is dissolved in the blood stream and then exhaled after a few minutes." This handbook concludes that a 0.5 μCi/cc concentration of tritium gas inhaled occupationally for 1 year would be the equivalent of a 5 rem dose. Therefore, the allowed level of tritium in $^3$He (via Spectra Gases' license—5 x $10^{-6}$ μCi/cc) would be equivalent to 5 x $10^{-5}$ rem if exposed occupationally to this concentration over an entire year (assumes 2400 $m^3$ of gas at that concentration was inhaled--much more than ever feasibly inhaled for medical imaging). A typical value of a very large dose of $^3$He (30 L, or 3 L per session for 10 sessions) would result in a dose of 6.3 x $10^{-10}$ rem.

The average American is subjected to multiple sources of radiation exposure, many of them from natural sources. By far the largest natural source of radiation exposure to humans is from radon (200 mrem/yr). However, even the chemical makeup of the human body includes carbon-14 and potassium-40

(40 mrem/yr). Cosmic rays deliver a continuous natural source of ionizing radiation (27 mrem/yr). Because the dose calculated above for tritium in $^{3}$He is a roughly billion times smaller than the one year burden from these natural exposures, we argue that its contribution constitutes an insignificant risk.

On behalf of the Hyperpolarized Media Study Group of the International Society of Magnetic Resonance in Medicine,

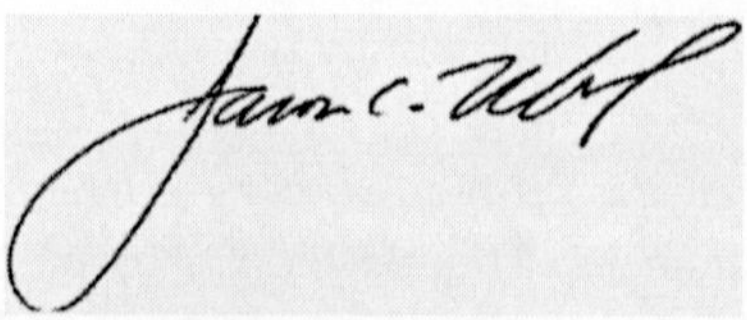

Jason C. Woods, Ph.D.
Program Chair, Hyperpolarized Media Study Group
Assistant Professor of Radiology, Physics, & Molecular Biophysics
Assistant Dean of Arts & Sciences
Washington University

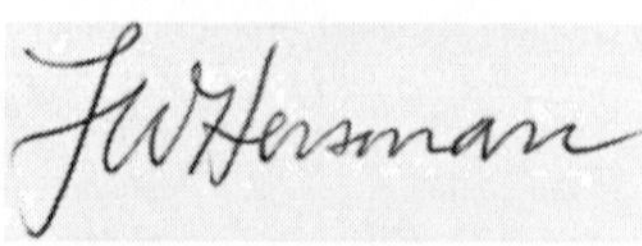

F. William Hersman, Ph.D.
President, Hyperpolarized Media Study Group
Professor of Physics, University of New Hampshire
CEO, Xemed, LLC

Jason C. Woods, Ph.D.
Assistant Professor of Radiology, Physics, & Molecular Biophysics
Assistant Dean of Arts & Sciences
Director, MARC uSTAR Program

Division of Nuclear Materials Safety
U.S. Nuclear Regulatory Commission
Cc: Spectra Gases, Inc. (Jack Faught, Keith Darabos)
3434 Route 22 West
Branchburg, NJ 08876

November 18, 2009

RE: release of $^{3}$He gas

To whom it may concern:

I write in support of the release of $^{3}$He gas for medical imaging to Dr. Giles Santyr and Dr. Grace Parraga, of the Robarts Research Institute in London, Ontario. Both Dr. Santyr and Dr. Parraga are pillars of the hyperpolarized gas community. They have active collaborations within the United States, are both at the cutting edge of medical-imaging research; their requests for gas release should be granted.

The pulmonary-imaging group at Robarts has broken new ground in the quantification of ventilation defects and their relationship to disease severity and progression in both asthma and COPD. Researchers and clinicians in the United States have directly benefited from their work and will continue to do so if this request is granted. Their research is mainly funded by the Canadian Institutes of Health (similar to our National Institutes of Health), and they have advisers and collaborations with researchers in the United States. I personally have visited their facility in the past 6 months, in part to discuss a collaboration with our group at Washington University and in part in an advisory role, as I serve on an external board of advisors to one of their CIHR grants.

In addition to being active contributors to the US and international community of medical imaging, they have developed good educational-industrial partnerships with US companies such as General Electric. Their past work with US companies has resulted in significant improvements of ultrasound, CT, and MRI technology, which benefits the US companies, US researchers, and US citizens at large. A failure to release $^{3}$He gas to this research team would significantly impede progress in the field and would have a detrimental impact on collaborations with US researchers and corporations.

Sincerely,

Jason C. Woods, Ph.D.

Jason C. Woods, Ph.D.
Assistant Professor of Radiology, Physics, & Molecular Biophysics
Assistant Dean of Arts & Sciences
Director, MARC uSTAR Program

Division of Nuclear Materials Safety
U.S. Nuclear Regulatory Commission

Cc: Spectra Gases, Inc. (Jack Faught, Keith Darabos)
3434 Route 22 West
Branchburg, NJ 08876

November 19, 2009
RE: release of $^3$He gas
To whom it may concern:

I write in support of the release of $^3$He gas for medical imaging to Dr. Frank Thien, of Monash University & Box Hill Hospital in Melbourne, Australia. Dr. Thien has recently been successful in performing $^3$He MRI with gas transported via an intercontinental commercial airline. Their work supports the success of a model of central production and distribution, even to a destination halfway around the earth from the production facility.

Dr. Thien is funded by the National Health and Medical Research Council, which is the Australian equivalent of our National Institutes of Health. He has active collaborations in Europe and in the United States with Dr. Kim Prisk at the University of California San Diego. Their collective work together is very scientifically productive and important to the field of hyperpolarized-gas imaging. Researchers and clinicians in the United States have directly benefited from their work and will continue to do so if this request is granted.

In addition, the amount requested (30 STP liters) is rather modest and will represent only a very small fraction of the total medical-gas utilization of $^3$He.

Sincerely,

Jason C. Woods, Ph.D.
Program Director, ISMRM Hyperpolarized Media Study Group

# CHAPTER SOURCES

Chapter 1 - This is an edited, reformatted and augmented version of a Congressional Research Service publication, R41419, dated December 22, 2010.

Chapter 2 - This is an edited, reformatted and augmented version of a US Government Accountability Office Publication, GAO-11-472, dated May, 2011.

Chapter 3 - This is an edited, reformatted and augmented version of a U.S. International Trade Commission publication, ITS-02, dated June, 2009.

Chapter 4 - This is an edited, reformatted and augmented version of the Opening Statement of Chairman Brad Miller before the US House of Representatives, Committee on Science and Technology, Subcommittee on Investigations and Oversight, dated April 22, 2010.

Chapter 5 - This is an edited, reformatted and augmented version of the Opening Statement of Dr. William K. Hagan before the US House of Representatives, Committee on Science and Technology, Subcommittee on Investigations and Oversight, dated April 22, 2010.

Chapter 6 - This is an edited, reformatted and augmented version of the Statement of Dr. William F. Brinkman before the US House of Representatives, Committee on Science and Technology, Subcommittee on Investigations and Oversight, dated April 22, 2010.

Chapter 7 - This is an edited, reformatted and augmented version of the Testimony of Thomas R. Anderson before the US House of Representatives, Committee on Science and Technology, Subcommittee on Investigations and Oversight, dated April 22, 2010.

Chapter 8 - This is an edited, reformatted and augmented version of the Testimony of Richard L. Arsenault before the US House of Representatives, Committee on Science and Technology, Subcommittee on Investigations and Oversight, dated April 22, 2010.

Chapter 9 - This is an edited, reformatted and augmented version of the Testimony of W.P. Halperin before the US House of Representatives, Committee on Science and Technology, Subcommittee on Investigations and Oversight, dated April 22, 2010.

Chapter 10 - This is an edited, reformatted and augmented version of the Statement of Jason C. Woods, Ph.D. before the US House of Representatives, Committee on Science and Technology, Subcommittee on Investigations and Oversight, dated April 22, 2010.

# INDEX

## A



## B



## C

## D

## I



## J



## L



## M

## N



## O

## S



## T

## U



## V



## W



## X



## Y